HOUSEHOLD WASTE IN SOCIAL PERSPECTIVE

Household Waste in Social Perspective
Values, attitudes, situation and behaviour

STEWART BARR
University of Exeter

Ashgate

© Stewart Barr 2002

Published by
Ashgate Publishing Limited
Gower House
Croft Road
Aldershot
Hampshire GU11 3HR
England

Ashgate Publishing Company
131 Main Street
Burlington, VT 05401-5600 USA

Ashgate website: http://www.ashgate.com

British Library Cataloguing in Publication Data
Barr, Stewart
    Household waste in social perspective. - (Ashgate studies
    in environmental policy and practice)
    1. Refuse and refuse disposal - Social aspects
    2. Environmental responsibility
    I. Title
    363.7'288

Library of Congress Control Number: 2002100074

ISBN 0 7546 1918 4

Printed and bound in Great Britain by
Antony Rowe Ltd, Chippenham, Wiltshire

Contents

List of Figures

List of Tables

Preface

The growing awareness that environmental problems have more to do with the decisions of ordinary people as opposed to governments and corporations is the rationale for this text. More significantly though, the realisation that the resolution of environmental problems involves not merely technical or economic adjustments, but real shifts in attitudes and behaviours of all people, is the core message of this book. For decades governments at various levels have attempted to deal with environmental ills by arguing that technical progress will eliminate the practices that blight the natural environment that we have inherited. Yet environmental degradation is now moving at a far faster rate than technological advances can hope to keep pace with. If technological progress cannot save the planet, then perhaps economic incentives or penalties might change the human behaviours that have led to such a polluted environment. Unfortunately, with society moving towards enrichment and a greater appreciation of personal wealth, taxation on commodities such as motor fuel or energy meet with stiff public opposition.

These fundamentally technocentric arguments have now become outdated and fail to take into account the fundamental cause of environmental problems. Lying at the heart of global warming, water shortages, resources exploitation and a growing waste mountain are the decisions that each individual on the earth takes. These decisions can of course be altered by technological advances or economic adjustments, but once these options have become exhausted by public acceptance or the pace of environmental degradation, it becomes apparent that only a shift in attitudes towards how we live our lives and the actions we take every day will actually have a lasting effect on the environment. Attitudes cannot be changed by what are seen as unfair taxes and certainly cannot be altered by technology that makes it easier for us to live our over-exuberant lives.

Yet we know surprisingly little about why people act in the way that they do. This is especially true when the five fundamental environmental behaviours are examined (water saving, energy conservation, transport use, green buying and waste management). Previous work into such behaviours has focused either on a very specific behavioural area and has paid attention to pre-defined models of how individuals act. Alternatively, past research has examined the rhetorics and communicative aspects of environmentalism in a qualitative genre. It is argued here that a new approach in the study of environmental behaviour is required. This should above all be holistic and use ideas from all disciplines. It should also be rigorous and produce useful data that are of utility to policy makers.

Using waste management as a vehicle for this approach, this book sets out to provide the basis for a new framework of conceptualising and examining environmental behaviour. This begins with an extensive review of the literature concerning environmental behaviour and then proceeds to develop an innovative framework for conceptualising such behaviour. Through the use of an extensive case study, the book demonstrates the utility of the approach taken in order to show how useful data concerning the influences on waste management behaviour can be researched and used to formulate policy.

Overall, the book aims to break new ground in a search for a sustainable future for our planet which is grounded in the recognition of personal agency to effect change.

Stewart Barr
Exeter, September 2001

Acknowledgements

This text has arisen out of research which has taken the best part of three years to conclude. As a consequence, I would like to thank the Department of Geography for their financial assistance, in the form of a full PhD scholarship, which has made this work possible. I should also like to thank Mr. Andrew W. Gilg and Dr. Nicholas J. Ford for their continual support and advice throughout every stage of my research. Their guidance has proved invaluable in the planning, implementation and write-up of the research. I should also like to thank other colleagues in the Department of Geography for advice and support, in particular Mr. Mike Leyshon who has provided an invaluable sounding board for numerous ideas in the research. I would also like to thank Professor Stephen Lea (School of Psychology) and Dr. Mark Brewer (School of Mathematical Sciences) for their advice concerning the statistical analysis in this book. Finally, my thanks must go to Professor Paul Selman (University of Gloucestershire) and Professor Gareth Shaw (University of Exeter) for deeming the work on which this text is based worthy as suitable for a PhD degree. Their comments and encouragement are gratefully acknowledged.

I would also like to thank those colleagues in academic departments away from Exeter. In particular, Mr. Ian Bailey of the Department of Geographical Sciences at the University of Plymouth who provided valuable advice and guidance with the drafting of the cover letter and questionnaire used in the empirical part of this research.

The continued financial and emotional support offered by my family and in particular my father, Mr. Robert Barr, who has been of great utility during all stages of the research.

I would also pay tribute to some of my closest and by now most tolerant friends who have endured exposure to some of my most extreme thoughts and ideas. My friends at Sidwell Street Methodist Church in Exeter have provided some of the most valuable assistance, in the form of fellowship and excellent food. In particular, I would like to thank Mr. Albert Fogarty, Mrs. Margaret Fogarty and Mr. Edward Yeates who have made invaluable comments on the early stages of the text.

Overall, I would like to thank all those I have come into contact with over the past two and a half years, who in some way have helped me to formulate ideas and often just to get away from what is seemingly an endless task. Inasmuch, I would like to thank Jesus Christ, without whose guiding light I would be lost.

1 Introduction: the waste problem in social perspective

Introduction

This book focuses on the issue of household waste management and how understanding individual attitudes and behaviours can make a significant contribution to an appreciation of what factors shape behaviour and how these might be changed. Waste is, of course, only one environmental behaviour where this can be appreciated. As the forthcoming paragraphs demonstrate, the world faces a plethora of inter-related environmental problems. The programme that has been institutionalised at the highest governmental level globally is 'sustainable development'. This ambitious project, integrating environmental, social and economic goals, faces many challenges, from basic definition to practical implementation. Yet the underlying challenge facing academics and policy makers today is not necessarily an environmental or fiscal one, but social. Until the majority of people change their attitudes towards the environment and sustainability and, until their behaviours change as well, the search for a sustainable future for the earth will elude us. The prospect of using a social understanding of environmental problems is one that has the potential to make the realisation of environmental sustainability more tangible than ever before. First, however, sustainable development and its challenges must be fully placed in context.

The Environment, Sustainable Development and Citizens

One of the most important academic and social issues of the late twentieth century is the reluctance of humankind to respond decisively to growing evidence of resource depletion and environmental deterioration (Munton, 1997, p.147).

This decisive statement on the state of human attitudes towards the environment is just one of many commentaries that have argued that unless hard choices about the environment are made now, the future of life on earth will be in doubt. Environmental problems are well documented in almost every sphere of the natural and human world. Global warming, ozone depletion, forest degradation, desertification, eutrophication, acidification, the litany continues. The consequences of these purely 'environmental' problems have significant social and economic impacts. As an example, global warming is already thought to have had serious effects worldwide, ranging from flooding due to enhanced rainfall (for example in Mozambique and India in 2000), to a particularly harsh El Nino in 1998-99 in Western America, to severe drought conditions in the Sudan in 2000. These 'natural' events, for example flooding, have familiar catastrophic results. Industrial shutdown, crop destruction, communications failure all result in economic disturbance as a consequence of flooding. Disease, starvation and in the long term increased poverty constitute some of the social costs of flooding. Just as the effects of global warming are diverse (flooding, storms and drought) so are the causes. Human activities produce greenhouse gases from numerous sources, including industrial plants, power stations, cars, aeroplanes, and landfilling of biodegradable waste.

The salient point to emphasise here is that environmental problems cannot be seen in isolation. They have human causes, human effects and are often the result of not merely of human economic mismanagement, but human poverty, leading to further degradation of the environment. Recognition of this fact has been late in reaching the governmental elite. However, the nations of the earth are now embarked on a programme that will hopefully begin to address these complex and acute problems. As outlined in the following sections this programme, called 'sustainable development', is now 'the message' for environmental protection, economic development and social progress.

In the following sections, the origins of sustainable development as a concept are examined initially, after which the conflicts regarding its definition are assessed, followed by an analysis of the practical working of sustainable development. Finally, an examination of the role of citizens within sustainable development is considered.

The Origins of 'Sustainable Development'

O'Riordan (1993) reminds us that the notion of 'sustainability' is an historic 'ideal' that has existed for thousands of years. What may be termed more accurately as 'subsistence' or 'stewardship' is still a norm for a considerable number of the world's population in certain inhospitable areas. However, the processes of modernisation, enrichment and, more recently, car use has meant that the Western world has become very much the antipathy of sustainability (Gatersleben and Vlek, 1998). Stark realisation of this problem began in the 1960s with the doomsday scenarios of Carson (1962) and Meadow's *et al.* (1972) predicting environmental catastrophe. Although the spectres of immediate environmental catastrophe and definable natural limits to growth, which first came to the fore in the 1960s, have subsided, a more holistic approach to the environment has emerged. Several conferences have been held at a global level since the Stockholm meeting in the early 1970s which have sought to bring the environment to the world stage (e.g. World Commission on Environment and Development, 1987; United Nations Conference on Environment and Development, 1992). Throughout these fora and many more besides, the concept of sustainable development has been 'reborn' and is now seen as *the* way forward in global development.

Sustainable Development: Definition, Meaning and Conflict

In 1987 the Brundtland Commission (WCED, 1987, p.43) defined sustainable development as '...development that meets the needs of the present without compromising the ability of future generations to meet their own needs'. Such a broad statement might be seen as so ambiguous as to offer nothing to the plethora of problems facing both the environmental and social world today. Indeed, O'Riordan (1989, p.93) describes the concept as being the '..refuge of the environmentally perplexed'. Nevertheless, O'Riordan (1993, p.37) does have to concede that 'No public figure or private corporation can afford to speak any other language'. This, in essence, is the key point, as will be demonstrated below - namely that although the vagueness of the concept is a difficulty, the wide-ranging and all-encompassing nature of the concept allows the idea to be sold and marketed far more easily to a sceptical public. Thus Wilbanks (1994) has argued that this has been the success of sustainable development (SD), namely that ambiguity and integration are seen as key virtues, not barriers.

However, a reading of the extensive literature on sustainable development reveals a sharp contrast between the 'marketable' and 'non-marketable' aspects of sustainable development. What Gibbs *et al.* (1998), among many more, have termed 'weak' and 'strong' sustainability presents the widest gap between the marketable and non-marketable approaches, respectively. There are many aspects to both of these approaches which cannot be detailed here, but those who support the weak approach are generally seen as technocentric (anthropocentric) and market-orientated whilst supporters of the strong position are usually ecocentric and intervention-orientated (Pearce, 1993; Gibbs *et al.*, 1998). However, the principal division comes with regard to the concept of capital. Pearce *et al.* (1989) argue that proponents of the weak approach believe that as long as the same (or more) amount of capital is passed on to the next generation, then the goals of SD can said to have been met. The strong sustainability followers, however, would urge that certain 'critical' amounts of natural capital stock must be maintained for the next generation (e.g. Pearce *et al.*, 1989; Pearce, 1993; Turner, 1998). What Turner has defined as a 'safety' level for such capital is vital for a number of reasons outlined by Pearce *et al.* (1989). Primarily, current use of non-renewable resources is such, that when they expire, there will be no viable resource base with which to meet current energy requirements (non-sustainability). Secondly, the presumption that new resources will always be found is not without doubt (uncertainty). Thirdly, following the path of only preserving capital stock which is economically viable means there is the risk of 'irreversibility'. Finally, degradation of natural capital means that the poor will be adversely affected, a point totally at odds with one of the key elements of SD - equity.

Of course it should not be forgotten that SD is not just about sustainability, although that is often held to be its rationale. 'Development' is a key aspect which is often forgotten. Pearce *et al.* (1989) have argued that 'development' is a value-laden word which has many facets. In this context SD contains three principal elements. First, a value of the natural, cultural and built environment. Second, a time horizon of years, decades and centuries. Third, both an intergenerational and intragenerational component. Within this framework, various words are casually incorporated such as equity, equality, peace, etc. Yet the volume of literature surrounding the concept of sustainable development is so vast that difficulties of scope and definition arise frequently. Wilbanks (1994) acknowledges this problem and details several conflicts that SD has tried to resolve, but in reality will take more than warm words to settle. Primarily, the old problem of 'conservation vs. growth' still has relevance today. Many misunderstand that SD is about development, of which growth is a part. However, most people still perceive the two as indistinguishable. Although most proponents argue that development can occur with conservation, this is a vexed question. A further problem comes with the democratic process, i.e. who decides on SD policies, the people or the government. A considerable literature has emerged concerning the inclusivity of sustainable development and the incorporation of thus far marginalised interest groups. Yet the degree to which this is practically possible is strongly debated. On the one hand, *Agenda 21* (UNCED, 1992) states that dialogue about local sustainability policies must be undertaken with all citizens, yet on the ground, decisions often have to be made that are unpopular. Indeed, can dialogue be maintained with people who are not interested? This leaves the vexed question of whether those who aren't interested are either forced to participate or left behind, which brings one back to an elitist position. This relates to a third issue of whether hard choices are going to be made or whether tinkering with the environment is sufficient. Most environmentalists and those in political opposition would opt for the former, whilst governments, for reasons of electoral viability, opt for the latter.

Sustainable Development in Practice

Implementing sustainable development in practice is a key difficulty for policy makers who are faced with the unenviable task of selling SD to a sceptical and apathetic public. Keeping the environment and sustainable development on the political agenda has thus proved rather a difficult task. In a rather over-zealous and optimistic discussion concerning the environment and sustainability, O'Riordan (1997a) celebrates the arrival of New Labour in 1997 and its new 'radicalism'. However, hope of such groundbreaking environmental reform was shattered, as the environment as an issue was pushed further down the legislative ladder. In a more depressing tone O'Riordan (1997b, p.2) a few months later, is left to criticise the lack of attention to sustainable development by New Labour and details the cabinet split on the environment, leading him to postulate that:

> The overwhelming temptation will be to tinker with it, and trust that a few positive initiatives will keep the NGOs happy (O'Riordan, 1997b, p.2).

The situation 'on the ground' in local authority areas is little better. The fusing of the economic and environmental goals of sustainable development are critical if Local Agenda 21 (LA21) is to be successful. The problems are given full attention by a range of authors (e.g. Blowers, 1993; Evans, 1995; Gibbs *et al.*, 1996; Gibbs *et al.*, 1998, Rydin, 1998 Selman, 1996; Selman, 1998; Selman and Parker, 1997). Essentially, the underlying difficulty is that of 'change'. Local authorities have been reluctant to alter their practices for consulting and implementing environmental policy, as well as the way in which they treat issues in a holistic manner. There are a number of elements to this problem. First, different perceptions within and between local authorities about what constitutes sustainable development, lead to confused messages both within and emanating from local authorities. Second, there has been a tendency to place Local Agenda 21 officers in planning and environment departments, rather than economic development departments as well. Third, sustainable development policies tend to be in very traditional areas of governance such as waste recycling and land-use planning. Emphasis on economic development and social progress within a policy framework linked to environmental initiatives has achieved only grudging support. Fourth, the political unwillingness and apathy in local government for new initiatives, such as Local Agenda 21 (LA21), ensures that only in certain local authorities is the issue of sustainable development high on the agenda.

Citizens and Sustainability

Of course, although these problems appear severe, it must be borne in mind that this is still the early stage of LA21 and there is time to change, but this will depend on political will and more importantly, the will of the populace. Generating interest and action from individual people is probably the most important and difficult aspect of sustainable development on the ground. There is only now recognition that much of what was decided at Rio can only become reality if people, all people, change their attitudes and behaviours in many aspects of their life. Local Agenda 21 and its associated policies need support from citizens to make them work. The notion of 'Think Globally, Act Locally' posits that small actions for sustainable development will have significant impacts if everyone is involved. As Paul Selman has argued 'The active environmental citizen is pivotal to the process of sustainability' (Selman, 1998, p.180). Yet why do some people act and others not, especially when both groups might state that they are equally concerned? Indeed, how can attitudes and behaviour be changed to achieve the goals of sustainability?

As will be outlined in more detail below, the search for a 'human' solution to some of our most pressing environmental concerns has been the focus of relatively few academics and researchers. The role of government in the process of regulation and exhortation in economic terms has been seen as the conventional and accepted method by which to resolve environmental issues. Yet many environmental problems are the result of what appear to be quite insignificant actions by individuals. The amount of carbon dioxide produced each year by private motor vehicles is a major contributor to the total greenhouse gas emissions which are widely blamed for the recent rises in global temperature. Yet, there are no easily definable polluters in this case. Anyone who drives a car makes a small, but in cumulative terms, significant contribution to carbon dioxide emissions. The classic response to this situation has been, for example, to raise duty on fuel in order to discourage drivers from using their cars as often. Yet, such measures are socially unacceptable, as was demonstrated by the widespread fuel protests throughout Europe in September 2000. What such a response from the public indicates is that attitudes towards car use are significantly at odds with the need to reduce car use and subsequent pollution. It is unlikely, as the fuel protests demonstrated, that higher taxes would lead to a change in behaviour in the short term. Thus, behaviour change is only likely to result from tangible shifts in attitudes towards the environment within the context of a clear programme for sustainable development.

Municipal Waste and Environmental Sustainability

One specific area of sustainable development and a key aspect of Local Agenda 21 is the issue of waste generation, especially household waste production. This section outlines the problems of waste management in the United Kingdom at present and what actions are being taken by the Government to alleviate the problem.

Waste management is a growing concern in both the developed and developing world. The growth in population and financial wealth results in increases in the products that are both consumed and more crucially here, disposed of after use. Chapter 17 of *Agenda 21* sets out waste as a key problem in the search for sustainable development worldwide (United Nations, 1992). Continuing this theme, the European Fifth Environmental Action Programme (CEC, 1992) cites waste as a key arena for action. However, it is at the national and local levels that the waste problem is seen most vividly. In England and Wales, the amount of municipal waste that is produced per year is 28 million tonnes (DETR, 2000). This figure is currently rising at around 3% every year.

The problem of municipal waste is diverse. The term itself refers to '...household and other waste...' handled by local authorities in England and Wales (DETR, 1999a, p.13). In practice, the majority is from households and comprises both organic and non-organic material. The main sources of the former are kitchen and garden waste, whilst the latter comprises glass, paper, aluminium, steel, cardboard, plastics and so on. In the period 1998/1999, 25.1 million tonnes of waste arose from households. This represents 25 kilogrammes of waste per household produced each week, a rise of 1.5 kg on the period 1996/1997 (DETR, 2000).

Currently, the majority of this waste is landfilled (82%), with just 9.5% being recycled and the remainder being sent for incineration and energy recovery. These figures are variable across England and Wales, with some local authorities achieving over 40% recycling of household waste, whilst others as little as 1%. Indeed, the situational constraints placed upon the waste management industry mean that different forms of disposal are used in various parts of the country.

47 Is it all conditional /relative?

Key Players in the Waste System: Production to Disposal

Currently the nature of waste production can be seen as result of a linear process from the production of goods to their disposal (Blowers, 1993). In terms of household (or municipal) waste, the main producers of goods taken into the home are those manufacturers of durable and non-durable household appliances, furniture, packaging (including goods and food packaging) and miscellaneous consumer goods such as newspapers, writing paper and so on. In household terms, the origins of such products are diverse and the amount of material included in each product varies widely. There are no statutory regulations governing the actual amount of packaging given with an actual product (Bailey, 1999), although there is a non-binding code that encourages minimisation of packaging (DETR, 1998). However, whilst there is no restriction on the amount of packaging actually produced, there are regulations governing the recovery and recycling of packaging. By 2000, 52% of packaging per year had to be recovered, with 16% being recycled (DoE, 1997). Such regulations apply to companies with an annual turnover of £1 million or more.

Once products are produced, the role of the householder then becomes crucial. Householders can make a number of decisions regarding what to do with the products they have purchased and indeed if to purchase such products. Essentially, householders can choose to minimise their waste by not buying certain products and/or packaging. They may choose to reuse what goods they have bought. They may also choose to recycle used products by bring schemes (static recycling sites in public amenity areas) or kerbside schemes. However, the most favoured option is to dispose of the 'rubbish' produced into black plastic bin liners or grey wheeled bins. This waste is usually landfilled or occasionally incinerated.

Once these decisions are made, the waste is dealt with in different ways according to how the householder has sought to dispose of the waste. Local authorities mostly undertake any recycling, although there are growing voluntary composting schemes. Waste is collected by the Waste Collection Authority (WCA) and recycled either 'in house' or by the Waste Disposal Authority (WDA). The WDA ensures that all waste passed to it by the WCA is disposed of.

However, to encourage recycling, the WDA must pass on any savings made in the diversion of waste from landfill (for example, to recycling) to waste collection bodies, such as WCAs or voluntary organisations. Indeed, to reduce landfill use, landfill operators now charge a levy on every tonne of waste sent to landfill, with 20% of the revenue from this charge being put back into environmental organisations and the rest being paid to the Treasury (Rydin, 1998).

As can be seen, the process from production of consumables to disposal of these products in various ways involves business, householders, government and NGOs. Having now outlined the brief role of each stakeholder in the process, it is prudent to examine the future targets set for waste management and the vital role of householders therein.

The Waste Strategy and Government Targets

The UK Government finally published its *Waste Strategy 2000* (DETR, 2000) for England and Wales in June 2000 after a lengthy period of consultation (DETR, 1998) and draft proposals (DETR, 1999a). The strategy outlines the targets for waste management in the coming decades. The Government wishes to see that by 2005, 25% of municipal waste is recycled (the target originally set for national recycling by 2000 in the previous *Waste Strategy*, (DoE, 1995)), with 40% recovery of waste by the same date. By 2010, 30% should be recycled, with a total of 45% being recovered. Finally, by 2015, 33% should be recycled with 67% being recovered. The implementation of the European Union Landfill Directive (DETR, 1999b) meant that the UK Government was compelled to set tough targets on landfill, since the directive states that by 2020 biodegradable waste going to landfill must be at 35% of 1995 biodegradable waste arisings.

The principles that govern these targets are crucial to understanding *Waste Strategy 2000*. First, the principle of BPEO or 'Best Practicable Environmental Option' is used to assess waste management options and encompasses the need for the best outcome for the environment, at manageable cost, both in the short and long terms. Second, the 'Waste Hierarchy' stipulates the need to shift current waste trends away from landfill towards increased reliance on minimisation, reuse and recovery (recycling, composting and energy) and only using disposal as a last option. Figure 1.1 below shows the Waste Hierarchy.

Figure 1.1   The Waste Hierarchy (Source: DETR, 2000)

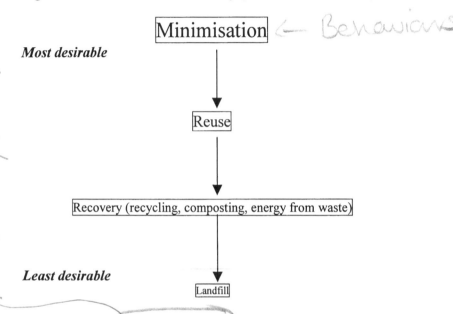

Most desirable

Minimisation ← *Behaviours?* (handwritten)

Reuse

Recovery (recycling, composting, energy from waste)

Least desirable

Landfill

(handwritten, left margin): dealt with near to ↗ creation point.

(handwritten, left margin): Sustainable ↗ option?

Third, the Proximity Principle suggests that waste should be dealt with as near to its point of creation as possible and is a key part of the Government's strategy for regional sustainability. In summary, the strategy sets out to reduce waste and deal with that waste which is produced on a local basis, in a cost effective, but environmentally sound manner.

The Role of Householders

Achieving these targets will not be simple and the Government has set out a number of initiatives in *Waste Strategy 2000* (DETR, 2000). The majority of these policies and initiatives are aimed at stimulating recycling and recovery from an economic perspective (see above). It has often been argued that until growth in the market for recyclate flourishes, there will be little increase in actual recycling rates. Creation of such a market would enable local recycling programmes to flourish. This assumes that there will be the recyclable material with which to undertake this activity. However, the major providers of such material are individuals, i.e. householders. Unless this group participate on a regular and steady basis, market

(handwritten, left margin): until more money is made from recycling it won't happen

stabilisation and growth will be difficult to achieve. Indeed, for *Waste Strategy 2000* to work in its desired form, people will need to buy less produce with packaging, reuse as much of this as they can and only throw away what can be recycled. In essence, a fundamental shift in attitudes and behaviour is required if the targets outlined above are to be attained. Thus, *Waste Strategy 2000* (p.51) concludes that:

> Individual consumers and households have a vital role to play in achieving sustainable waste management. We can all help by:
>
> • Buying products which will produce less waste and those made from recycled materials;
>
> • Separating our wastes for recycling and composting kitchen and garden waste;
>
> • Participating in local debates about how best to manage our waste.

Household Waste Behaviours

As the previous section has demonstrated, the householder has a crucial role to play within the waste system and the problem of waste production is only likely to be reduced if most households change their behaviour. We have, as yet, not specified in detail what such behaviours are. Figure 1.1 shows that such behaviours can be classified as Minimisation, reuse and recovery. There is some contention among experts regarding the use of these terms. Minimisation refers, for example, to the reduction of waste being produced in the first place and could equally be termed waste 'prevention'. Recovery has a number of different facets and, in environmental terms, variable acceptable processes by which to reproduce goods from waste. Without doubt the most popular form of 'green' waste management is recycling. For several decades local governments and environmental charities have advocated the industrial reprocessing of household waste as a way of returning some value from what has been used. Yet recycling is way down the waste hierarchy in comparison to preventing waste being produced in the first place or attempting to reuse products after they've been utilised once. It is therefore prudent to investigate what household waste behaviours can be used in order to manage our waste.

Minimising, preventing or reducing waste As stated above, stopping waste being produced is of crucial importance. But what does this mean in the household context? As was demonstrated above, there are opportunities for others to prevent potential waste from entering the system, but there are also ways that householders can prevent waste being produced. These focus mainly on shopping habits such as reducing the amount of produce bought with packaging and, for example, not using plastic bags for packaging fruit and vegetables. Taking a shopping bag rather than using plastic carrier bags also provides a useful means of waste prevention. There is also the opportunity to think ahead in the waste process by looking for packaging that can be reused easily, such as cake tins and margarine tubs, or looking for packaging that can be recycled with minimum effort. Overall, there are numerous opportunities for the householder to reduce the amount of waste and effort before products even enter the home.

Reuse   Once goods have been purchased and are present within the home there is the opportunity to use them again after their initial role has been fulfilled. Numerous types of container can be reused, such as margarine tubs, bottles, yoghurt pots, cardboard boxes and so on. Paper can also be reused for all kinds of purposes, such as shopping lists and reminders. Clothes, cloths and other fabrics can be washed and used for other tasks, such as dusters or other cleaning materials. Reuse, in this sense, is highly innovative. This can in one context be seen as a problem, since there may be social barriers to action, as will be seen below. However, in the positive sense, reuse offers a clean and cheap means by which to reduce the amount of waste being produced.

Recovery, recycling and composting   Recovery strictly covers three activities, but as the focus here is on household actions, energy recovery will not be alluded to here.

Recycling   The number of materials that can be recycled is large, but where these materials can be recycled is spatially variable. Hence, the often very visible logo on the side of many items of packaging showing a green arrow within a circle merely refers to the recyclability of that item of packaging and not whether it will be recycled. This is still the householder's choice. This being the case, individuals need to be constantly aware of what can be recycled in their local area and when, if applicable, to place their recycling receptacle out for collection. Nonetheless, if circumstances are favourable, then a large number of materials, from plastic bottles to steel cans, can be recycled into the same or different products.

Composting   This method of waste recovery is growing in popularity and takes a number of forms. Householders can now purchase a number of containers that enable food and garden waste to decompose and form rich compost. Wormeries, green cones and other composters are now available from local authorities at reduced cost. Also taking some hold is the growth in community composting where garden waste is taken by householders to a communal collection point within the community and composted.

Overall, there are three fundamental areas of choice that the householder has at their disposal and these can cumulatively produce a situation where the amount of waste being passed from the household is significantly produced. → Choice of waste reduction methods.

Encouraging Action: An Evaluation of Waste Awareness Campaigns

Key to the UK Government's sustainable development strategy, outlined in 1999 in *A Better Quality of Life* (DETR, 1999c), is the encouragement of changes in social attitudes to various environmental problems. *Waste Strategy 2000* details a number of awareness initiatives aimed at increasing public knowledge of waste issues. The *Are You Doing Your Bit?* Campaign stresses how subtle changes in lifestyle can have a significant impact on the environment and waste is one of the key elements in this initiative. The techniques employed by the campaign and the messages therein are designed specifically to raise awareness concerning the human effect of wasteful lifestyles on the environment and an appreciation of the benefits of everyone 'doing their bit'. The campaign uses one of the most popular methods by which to disseminate such information – grand statistics on the state of the environment and the impact of any alleviating action.

The national charity for waste issues, *Wastewatch*, is spearheading the *National Waste Awareness Initiative* launched in autumn 2000. This ambitious campaign is actually more than it first appears since its role is not merely to inform the public of waste issues, but also to provide support for new legislation on waste management and encourage infrastructural improvements in waste facilities. However, as its name suggests, the principal domain of its focus will be on making people aware of the waste problem. Through a unified national campaign, which is intended to complement and aid other initiatives, the strategy is defined by 'hard hitting messages' regarding waste production and its consequences. Such messages are intended to be locally contextualised, although since the campaign is at an early stage, no literature is available on which to base any further assessment of the campaign. What can be said is that the initiative is implicitly awareness based, relying on hard hitting messages to shock people into action that, with a local focus, will change their attitudes and behaviour.

The DETR has also instituted a website for informing the public about the ways in which they can reduce waste (at www.useitagain.org.uk). Again, the site offers useful advice on how to reduce and reuse waste and aims to raise public awareness of how to tackle the waste problem. Such 'green lists' of appropriate actions to help the environment are a common feature of awareness-based campaigns, such as the *Going for Green* initiative and the *Action at Home* programme of *Global Action Plan*, a national environmental charity.

This brief review of campaigns has only touched on national initiatives for action and there are many more aimed at a local audience initiated by NGOs and local government. Awareness of an environmental problem and how to begin resolving that problem is evidently a key requisite for behaviour change. These campaigns all help to reinforce the important messages concerning the severity of the waste problem, the need to act in the short term and the core actions that can be undertaken to alleviate the problem.

The implicit assumption has, however, been that there is an intrinsic and linear relationship between awareness of a problem and its resolution, and behavioural commitment. Whilst there is no question that awareness is a vital factor in modifying behaviour, it is argued here that the relationship between awareness and action is complex and that numerous other factors can intervene in the relationship between knowledge and action.

Previous research in this area suggests that individuals face important barriers and motivating influences that impact on their willingness to undertake a given action and behave accordingly. As the research outlined in Chapters Two to Four demonstrates, although knowledge plays its part, there are numerous motivating or disabling factors that impact on peoples' opinions and actions. These range from fundamental environmental values about the intrinsic or extrinsic value of nature to the effect of other peoples' actions on individual behaviour.

It seems logical to understand why people do and do not partake in certain behaviours and what influence their attitudes have on those behaviours, before trying to change those attitudes and behaviours. There is consequently a need to address the question of why certain people may hold a positive attitude toward recycling, but don't act and, conversely, why those who seem quite uninterested in waste issues, reuse and recycle waste. It is argued here that such an approach to waste management and, in more general terms, environmental behaviour, offers a better prospect of changing fundamental attitudes and behaviours in the long term since a holistic range of factors that change behaviour are accounted for and not just awareness of the problem. Indeed, such an approach can offer policy makers, who are on the most part the users of awareness campaigns, an opportunity to develop innovative policies that relate to the actual determinants of behaviour in a specific area rather than policies that are aimed at a general swathe of the population and that have only a weak causal grounding in changing behaviour.

A New Approach to Understanding Environmental Behaviour

Understanding the reasons why people do and do not behave in given ways is an issue with which social scientists have grappled for considerable time. Previous research into environmental behaviour has been dominated by two distinct approaches. The first has been focused around a psychological understanding of behaviour that has been relatively uni-disciplinary and North American based. Researchers have focused on previously conceptualised psychological models of behaviour and applied them to various environmental actions. The approach is highly constrained conceptually and is statistical in the extreme. The second approach has been common within the United Kingdom and has focused on a qualitative understanding of rhetorical and communication elements of sustainable lifestyles. This approach uses a qualitative understanding of behaviour to argue that environmental behaviour must be seen in individual context and that generalisation through frameworks or models of behaviour is undesirable since individual choices are essentially highly personally and socially contextualised and therefore cannot be predicted.

It is argued here that two fundamental shifts are needed to broaden the disciplinary scope of the psychological approach and to use this vehicle to meet the challenges posed in the United Kingdom by those who see quantitative research as deterministic. First, this book proposes an approach based on a geographical understanding of environmental issues. As a discipline, geography is defined primarily by its interest in investigating the relationships between humans and their natural environment. The insights that can be brought from the geographer are considerable within an interdisciplinary context. Second, it is proposed that the widely dismissed use of quantitative techniques for investigating human behaviour are re-evaluated and advocated as a complement to the more qualitative approaches to investigating social action. These two 'new' approaches are now considered in more detail.

The Geographical Approach

> A student of geography inevitably questions the relevance of what he is studying. In particular he is concerned with the practical application of his geographical knowledge and expertise to problems that affect the lives of people (Dawson and Doornkamp, 1973, p.1).

Geography has long faced the conflict between the 'pure' and the 'applied'. This is still the case today and is likely to be so for the foreseeable future. Yet it is odd that a subject that is fundamentally grounded in environment-human relationships has not grasped the nettle of sustainability more vigorously. Geography should bridge the gap between theoretical rigour and practical necessity, providing a synthesis for academic and applied work. Yet such a 'Green' geography (O'Riordan, 1989) has been slow to emerge from the challenges of WCED (1987) and the Rio conference (UNCED, 1992). That is not to state that no work exists in this area. However, it is notable that a subject which is split between the physical and human sciences, has not sought to more readily utilise and place at its heart the interest that emerges from environment-people relationships. This is not to argue that developments in human geography have necessarily hampered this process. Developments such as the 'cultural turn' (Robinson, 1998) have provided as many opportunities for research into environmental geography as the 'Quantitative Revolution' of the 1960s did. Rather, there has been a tendency to let others deal with the big issues of political significance. Therefore, whilst environmental planners, psychologists and sociologists examine human dimensions of environmental change, the subject best suited to take a holistic and fresh view concerns itself with other issues, to the extent that colleagues question the extent to which environmental issues, such as waste management, are 'geography'. The contribution that geography can make is still important and Clayton's (1976, p.101) question posed twenty five years previously still has resonance:

> Perhaps geography can do without environmental studies; how long can environmental studies manage with so few geographers?

The geographer therefore has a key role in examining the relationships between environment and people and as such the means by which people deal with their waste, a significant environmental issue, is of importance to the student of geography. Yet caution should be shown when espousing the benefits of one discipline to the detriment of another. Without doubt the subject area that has made the most significant contribution in this area of research is psychology. Work in the United States in particular has shown the utility of psychological research in examining human dimensions of environmental change. Indeed, the techniques used within the study at the end of this book are based on considerable psychological understanding. However, what environmental psychologists and, to a certain extent, environmental sociologists, have overlooked is the need for holism in the debate. Psychologists use various theories to explain behaviour, but there is little cross over between theories. The geographer can, to a certain extent, be somewhat of a scavenger in this context. Although a notion of conceptualisation is crucial to the geographer, the constituents of that conceptualisation are of less significance and can therefore comprise what the geographer sees as the most effective and holistic approach, as opposed to the most theoretically rigorous methodology.

In summary, the geographer can offer both theoretical insights as well as provide a clear holistic framework in which to work.

The Methodological Approach

Geography has moved through a number of distinct methodological phases since World War Two (Cloke *et al.*, 1991; Robinson, 1998). The 'Quantitative Revolution' of the 1950s and 1960s, with its reliance on logical positivism, gave way to humanistic and Marxist Structuralist approaches, later developing to form the 'political economy' approaches of the 1980s. The subsequent crisis of representation of the late 1980s led to the 'cultural turn' which has epitomised the geographical scene in the 1990s (Robinson, 1998). Yet this does not imply the need to utilise any particular methodology on the grounds of fashion or current acceptability.

Work using the qualitative genre has been undertaken into environmental sustainability during the 1990s. Eden's (1993) work on environmental responsibility, Burgess *et al.*'s (1998) research on environmental communication and citizenship, and Machnaghten and Urry's (1998) work on the rhetorics of sustainability have all utilised forms of qualitative understanding, such as discourse analysis, in order to uncover public attitudes towards environmentalism and sustainability. This work is of great value and provides useful insights into the construction of the meanings and messages of sustainable development.

However, these studies and others like them have been specifically interested in the meaning of sustainable development as a whole and have not intended to examine behavioural responses to specific attitudes. The current research seeks to examine the attitudes of people towards a given problem (waste) and the extent to which these attitudes are reflected in (waste) behaviour. Indeed, it seeks to understand why aspirations do not always lead to action and to make a contribution to policy related debates on waste. Such goals cannot be met with use of the previously mentioned techniques since there is no way of measuring individual attitudes, expressed in textual or verbal terms, against any form of standard behaviour measure. Indeed, comparison of such measures between subjects would be difficult. There is also the need to make informed policy recommendations that can arguably only be made on the basis of 'hard' evidence relating to attitudes and behaviours. This is not to state that qualitative data cannot make valuable contributions to debates on environmental action, but rather that a study of attitudes and behaviour can best make recommendations on the basis of standardised data.

Given that the political economy approaches and Marxist Structuralist genres imply that personal agency is of little importance in shaping behaviour, there is the spectre of using quantitative techniques even though Robinson (1998, p.475) states that:

> ...one of the most strongly recurrent themes within human geography during the past three decades is the widely held view that scientific or quasi-scientific methodology is inappropriate for dealing with problems studied by human geographers.

Yet the most successful theoretical and practical attempts to examine attitudes and behaviours in the environmental realm has been with the use of quantitative techniques. The voluminous literature produced annually in the United States on aspects of individual environmental action, such as waste recycling, water saving and energy use is evidence that such research contributes to valuable debates on environmental behaviour. The use of conceptualisations of environmental behaviour and the measurement of human attitudes and behaviours on well-founded psychometric bases is long established in the United States. Utilisation of statistical techniques has enabled workers to examine the determinants of environmental action. In many cases such research makes important policy as well as theoretical contributions and although strongly grounded in social-psychological theory, is fundamentally interdisciplinary.

Yet concerns raised regarding quantitative generalisation (e.g. Eden, 1993) must be addressed. It is argued here that what quantitative approaches lose in data richness, they compensate by ensuring that the data produced are generalisable with care. Such data are not used to make definitive statements of fact. They are intended to track patterns and trends within large sample populations that can inform both theoretical and policy debates. They therefore provide an avenue to explore the formation of attitudes, the relation of these attitudes to behaviour and the ability to make general statements and propositions regarding environmental behaviour.

Overall, the approach advocated here attempts to provide the debate concerning the determinants of environmental behaviour and, in particular, waste behaviour, with a holistic framework that incorporates understanding from a number of disciplines. The approach also seeks to re-establish some of the unjustified loss in respect for quantitative techniques that have been summarily dismissed by qualitative researchers.

Values, Attitudes, Situation and Behaviour

The development of a more holistic approach that re-establishes confidence in the role of quantitative techniques needs to meet two requirements. First, all factors that might affect behaviour need to be considered equally and with rigour. Second, well-tested and established techniques within the human sciences need to be used and adapted for the data involved.

To meet the first of these requirements, an extensive review of the work into environmental behaviour has been undertaken covering literature from many disciplines and sub-disciplines, including geography, environmental psychology, behavioural psychology, environmental sociology and environmental planning. Studies have been reviewed that have used rigorous psychometric models on the one hand and ad hoc analysis on the other. It has been found that there are three fundamental sets of variables that are likely to influence environmental and, specifically, waste management behaviour. First, 'environmental values', an individual's orientation towards the value of nature and the environment, have been linked to behaviour. Second, 'situational' factors have been linked to behaviour, comprising an individual's personal circumstances regarding demographic position, access to key services and their awareness and experience of the relevant behaviour. Third, 'psychological' factors can be linked to environmental behaviour. These are individual perceptions about the behaviour in question, based for example on the social acceptance of the behaviour. These three sets of factors are addressed in Chapters Two to Four and form the basis of understanding for the holistic approach taken here.

The second requirement described above relates to the methodological approach on which any use of quantitative techniques is based. A great deal can be learnt from the rigorous analyses of human behaviour undertaken in all areas of psychology and in particular the extension of some of the statistical assumptions that are made concerning the data in question. Such extensions of data requirements are not fundamental in nature, but do mean that multivariate analyses can be performed where, for example, they may not have been deemed appropriate within the discipline of geography. Hence, although certain statistical rules may have been stretched slightly, the payoff is a set of results that can reveal far more than bivariate statistics can hope to. The widespread use of these multivariate techniques, such as regression analysis, logit and discriminant analysis within psychology provide a way of examining quantitative data that opens new avenues for geographers examining human behaviour. Indeed, the results of such analyses, in the form of path diagrams and 'percentage explanation' statistics provide a helpful guide to trends in the data, whilst recognising always that statistics are only a tool and not the 'truth'. This consideration is an important one since all social scientists should bear in mind that statistics offer quantitative inferences concerning hypothesised relationships that cannot be 'proven' in the true sense.

Structure of the Book

Having now placed the waste problem and the role of households into context, the second chapter will outline the first set of variables that have been linked to waste behaviour. An introduction to the notion of values, attitudes and concern grounds the reader in how fundamental values relate to behaviour. The chapter then outlines the scales that have been used to relate values to waste behaviour, specifically focusing on the 'New Environmental Paradigm'. Other scales are considered and their overall impact assessed.

Chapter Three outlines the second set of variables that have been linked to waste behaviour and begins with a brief introduction on how structural variables have an overall impact on action. The role of social context is examined (for example, access to kerbside recycling), along with basic socio-demographic influences (age, gender, family types, car access, house types, occupation, income, education, political affiliation), knowledge of both the environment and waste issues and finally previous behavioural experience of waste or other environmental behaviours. The problems of using such linking variables are discussed at the end of the chapter.

The third set of linking variables is outlined in Chapter Four, with a brief description of their nature and scope. The various theories and variables are discussed in turn, emphasising initially the theoretical and psychological justification for their inclusion and then examining the ways in which such variables have been used to explain variation in waste behaviour. The chapter ends with a discussion of these variables and in particular the diversity of their nature and application.

Chapter Five examines how the three sets of variables outlined in the previous chapters can be conceptualised into a workable framework of environmental and waste behaviour. Previous models of behaviours are examined and critiqued. The reader is then introduced to a popular model of social behaviour, the Theory of Reasoned Action (TRA). This is described and its application to environmental research is outlined. This logical framework is then used to construct a conceptualisation of environmental behaviour, the elements of which are the variables described in Chapters Three to Five. This new and innovative framework of environmental behaviour is promoted and its utility to waste management is discussed.

From the abstract conceptualisation of environmental behaviour presented above, Chapter Six seeks to demonstrate how such a framework can be employed in a practical study of waste behaviour. The area of investigation is discussed and the implementation of the framework cited above is described. The social bases of the sample are outlined and compared to appropriate Census data. The chapter then moves on to describe the specific elements of the framework given above, examining in particular waste behaviour, intentions, values, situational factors and psychological variables. Cluster analysis of these variables is described and these are classified. Thus by the conclusion, the reader will have a grounding in the behavioural, attitudinal, values, situational and psychological bases of the sample.

After grounding the readers' knowledge in the descriptive statistics of the sample, Chapter Seven uses progressively higher order relationships in order to assign importance to the variables outlined in the conceptual framework by initially using bivariate statistics to examine salient relationships. These are then developed, using aggregating techniques such as factor analysis and explanatory methods, such as regression analysis, in order to synthesise waste behaviour in Exeter. A concise conclusion will be given of the determinants of each type of waste behaviour.

In Chapter Eight, the practical case example of Exeter will then enable a coherent account of the current research in respect of existing theories of environmental and waste behaviour, in particular from North America. The similarities and differences will be examined and the promotion of the new framework will be emphasised. The implications for academics and policy makers will be examined. For academics, the promotion of the holistic framework will be emphasised, in particular the need for those already working in this field to take up the challenge of using such a conceptualisation. For policy makers, the need to treat waste as differential will be promoted.

Leading on from Chapter Eight, the final chapter will emphasise the importance of individuals in the waste process as key decision-makers in the production-disposal process. The utility of the social psychological approach will be emphasised and discussed in the context of previous research that has critiqued quantitative approaches. Finally, the chapter will end with a note on the future direction of waste policy in the UK.

2   The Social Psychological Perspective I: environmental values and attitudes

Introduction

The first set of factors that have been linked to environmental behaviour constitute the most fundamental bases of human subjectivity towards the environment and society in which we live. This chapter examines the wide ranging research into environmental values, attitudes and concern that has been undertaken and provides an account of the variability in research that exists in this area. Psychometric scales from the United States are examined in the first instance since these form the basis of social-psychological understanding of environmental values and behaviour. Context is then provided for environmental values within the United Kingdom, although this work has generally been outside the social-psychological arena. Finally, a cross-national evaluation of the impact of environmental values is given.

Values, Attitudes and Concern

Considerable early research into environmental behaviour attempted to link general environmental attitudes, values and concern to specific behaviours. Latterly, this has been undertaken in a more sophisticated manner. However, these studies have given mixed results regarding the link between values and action that might be more to do with measurement instrument construction than actual variable efficacy. As will become apparent, the wide use of environmental 'values' as a measurement instrument has ensured that different workers have utilised alternative terms to describe their instruments. These are grouped here for clarity. Indeed, there is considerable overlap. The term used here is 'values', referring to an underlying environmental orientation, based upon a personal worldview.

However, studies using this criterion have also termed their scales 'attitude' items (most popularly). Indeed, scales using these criteria have also been termed 'environmental concern' scales (e.g. Weigel and Weigel, 1978). This is the term also used for one item questions concerning 'yes/no' questions regarding environmental concern! For clarity, studies that have examined the relationship between environmental values, attitudes and concern are grouped here under the subjective term 'values', although it should be noted that this is only one of three possible terms that could have been chosen! The individual use of the terms used in particular studies is given for completeness, but it should be noted that they all gauge a similar, general concept in relation to the global environment.

Environmental Values Scales: the *New Environmental Paradigm* (NEP)

Perhaps the most widely known and tested measurement scale for environmental attitudes and action is the *New Environmental Paradigm* scale (NEP) developed by Riley Dunlap and Kent Van Liere in the late 1970s (Dunlap and Van Liere, 1978). Dunlap and Van Liere argued that the NEP represented a shift in societal values away from the *Dominant Social Paradigm* (DSP) characterised by over-consumption and materialism, to the notion of the NEP, encapsulated by 'limits to growth' and the 'spaceship earth' concepts. Dunlap and Van Liere formed their twelve-point scale from a larger set of scale items and tested the validity of the scale upon residents of Washington State, United States. They not only found that there was higher than expected agreement with the NEP, but also that, as expected, those in environmental organisations scored more highly on the scale than the general public. However, whilst the scale provided good predictive validity for other attitudinal concepts, Dunlap and Van Liere reported that the attitude-behaviour correlation was weak, albeit significant. Hence, whilst agreement between the scales was reached on items that predicted acceptance of more stringent environmental policy, actual action was not in such high agreement.

Testing of the NEP and its relationship with behavioural actions has been undertaken by a number of researchers up to the present. However, before detailing these studies, it should be noted that concerns have been raised about the use of the NEP as a single scale. Albrecht *et al.* (1982) found that the unidimensionality of the NEP claimed by Dunlap and Van Liere was in fact not the case in their study of farmers and urbanites in

Iowa, United States. Through factor analysis Albrecht *et al.* derived three principal factors that constituted a multidimensional scale. These were broadly domains referring to the 'balance of nature', 'limits to growth' and 'man over nature'. Kuhn and Jackson (1989) also reported a similar result in their research on the NEP.

In terms of testing the relationship between the NEP and behavioural differences, Steel's (1996) analysis of the NEP in relation to behavioural indicators from an American study of environmental attitudes and behaviours indicated that there was moderate, but significant correlation between agreement with the six-item subscale NEP that he used and an index of environmental behaviours. This led him to the rather bold conclusion that:

> Attitude intensity in support of the NEP predicts self-reported environmental behavior and participation in environmental issues (Steel, 1996, p. 34).

However, Vining and Ebreo's (1992) longitudinal study of residents in Champaign, Illinois, United States found that, as reported by Dunlap and Van Liere over a decade earlier, scores on the scale were generally high across their sample. In Vining and Ebreo's case, they found that recyclers tended to have marginally higher scores on the three-factor NEP (used by Albrecht *et al.*, 1982) than non-recyclers. Nevertheless, as they point out:

> These results indicate that the differences between the recyclers and nonrecyclers, at least in terms of attitude, are a matter of agreement and not a contrast in fundamental values (Vining and Ebreo, 1992, p. 1603).

This is a sentiment expressed by Scott and Willits' (1994) study of Pennsylvania residents. Again, there was high overall agreement with their two-factor NEP, but relatively low attitude-behaviour correlations (not exceeding an *r* of 0.21). However, they argued that rather than the difference between oral aspiration and behavioural action being due to scale inefficiency, the disparity between attitude and behaviour might be 'real'. They went on to state a reason for this apparent anomaly:

> Given the amount of media coverage devoted to environmental problems, it could be that many people have learned the language of environmentalism without developing a simultaneous behavioural commitment (Scott and Willits, 1994, p. 255).

This appeared to be the case more vividly in Sweden where Widegren's (1998) study of environmental behaviour demonstrated not only poor correlative links between their six item NEP and pro-environmental behaviour, but also the fact that, as was found to the contrary in Dunlap and Van Liere's research in 1978, the general public scored more highly on the scale than members of an environmental organisation (in this case members of the Swedish Conservation Society or SNF). Such a finding contradicted the assumptions of the shift from the DSP to NEP predicted by Dunlap and Van Liere. Although Widegren acknowledges that the NEP was probably tapping high levels of environmental awareness that are widespread in Sweden, he argues that, as with Scott and Willits (1994) above, the importance of environmental values having an impact upon actual behaviour was slight.

Of course, it is often difficult to know exactly whether the study instruments' instability accounts for low correlation coefficients, or whether the low-level relationships reflect genuine attitude-behaviour inconsistency. However, although these studies show weak bivariate correlations between the NEP and behaviour, it must be remembered that, as Fishbein and Ajzen (1975) have reminded us, measuring attitude at the general (environmental) level and expecting behavioural correlation at the specific level is at best optimistic. Roberts and Bacon (1997) also supported this in their detailed analysis of consumer attitudes and behaviour. What the path model of Widegren (1998) did show was that the NEP had as important an indirect effect on behaviour as it did a direct impact. Hence, as seems logical, it might be that the NEP represents an underlying value structure that cannot, by its very nature as a global measurement instrument, be replicated in specific environmental actions. However, it is likely that it, or some other adequately formed instrument (see following section) could indirectly predict behaviour.

Other Environmental Values Scales

Indicative of the pioneering attitude research of the 1970s and no doubt related to the catastrophist theories of Meadows *et al.* (1972), were workers such as Maloney and Ward (1973) whose zealous article in the *American Psychologist* stated 'Ecology: let's hear it from the people'. Their attitude scale consisted of 34 items tapping individual attitudes towards environmental issues. In addition, they measured verbal commitment to act and actual commitment (behaviour), yielding a total of 134 items! Three groups were used to test the scales, members of the Sierra Club (a commonly studied environmental group in the United States), university students and lay people. The California-based study found that, in the words of Maloney and Ward (1973, p. 585):

> In colloquial terms, most people say they are willing to do a great deal to help curb pollution problems and are fairly emotional about it, but, in fact, they actually do fairly little and know even less.

Although attempts to shorten the scale were successful in reliability terms, the relationship between attitude and behaviour was not significantly improved and once again there is the salient question of whether multivariate analysis would have yielded more meaningful results, or whether the apparent lack of predictive power of environmental attitudes (direct or indirect) was indeed true.

A more successful attempt to relate general environmental attitudes to behaviour was given by Weigel and Weigel (1978). Their sixteen item scale (termed 'Environmental Concern' or EC), originally drawn from a pool of 31, tapped general values concerning the environment. It was compared to three measures of environmental behaviour in a medium-sized "Western" city in the United States. After administering the EC scale, the same respondents were asked to partake in a 'litter pick' and sign a petition and be part of a recycling programme. Weigel and Weigel found a correlation of 0.62 between their comprehensive behaviour index and the EC scale. This is certainly a good result and is better than those correlations reported for the NEP above. However, subsequent testing of this scale has urged caution to be exercised when evaluating this direct result (Tarrant and Cordell, 1997).

Another set of positive correlations between environmental attitudes and action was achieved by Thompson and Barton (1994) in their analysis of scales measuring ecocentrism, anthropocentrism, apathy and conservation behaviour in California, United States. Using multiple regression techniques, they found that in their first sample (general public), ecocentrism significantly predicted conservation behaviour, whilst anthropocentrism negatively did so. In their second study (college students) they found a similar relationship between ecocentrism and behaviour, but not anthropocentrism and action. They argue that this is likely to be due to the different populations being assessed, a point raised by Tarrant and Cordell (1997) (see Section 2.3.3 below). Nevertheless, their study demonstrates that their ecocentrism scale had more predictive value when assessing environmental behaviour than the previously mentioned EC scale of Weigel and Weigel (1978). In conclusion they state that:

> Ecocentrism appeared to tap a disposition toward environmental issues that was not captured in traditional measures of environmental attitudes with no ecocentric-anthropocentric distinction (Thompson and Barton, 1994, p. 156).

Having said this, it must be noted that the amount of explanation offered by the scales was still only (R^2) 50% and the addition of other variables to the regression model might have reduced or even excluded the significant variables from this analysis. Nevertheless, it does suggest some link between environmental values and behaviour that has some direct and/or indirect predictive value.

This is contradicted by findings of those such as Arbutnot (1977) in his analysis of recycling among two distinct population groups in Athens, Ohio. He distinguished between community recyclers and conservative churchgoers, the former representing ecologically active persons, the latter not. Using factor analysis to aggregate his survey items measuring attitudinal and value-based constructs, he found that whilst there was a difference according to 'environmental cynicism' (what is termed below as 'response efficacy', see Section 2.5 below) and 'ecological responsibility', there was no difference according to environmental attitude between the recyclers and church goers.

This finding is explored in more complex analyses of environmental behaviour and attitudes that have been undertaken into recycling by Oskamp *et al.* (1991) (see Oskamp, 1995b for a shorter resume) and green buying by Mainieri *et al.* (1997). Oskamp *et al.* (1991) studied kerbside recycling in California and used bivariate correlation as well as hierarchical multiple regression in order to elucidate the underlying variance in behaviour. Initially they found positive relationships between their 'pro-ecology' attitudes, but regression analysis revealed that when these attitudes were put into the context of the other variables they were assessing, the relationship was significantly negative. Mainieri *et al.* (1997) surveyed Los Angeles residents concerning their attitudes to green consumerism, actual action and a number of other demographic and psychological traits. As was hypothesised above, when more variables were added into their regression model, the direct predictive value associated with their aggregated attitude index fell in relation to other variables. Their aggregated scale consisted of the NEP, EC and other environmental attitude items. In general, Mainieri *et al*. (1997, pp. 201-202) state that:

> 'On the average, environmental concern among the respondents was moderate to strong, but this concern did not usually carry over to their environmental buying habits and participation in environmental behaviour' and that 'Conceivable reasons that the respodents' proenvironment consumerism lagged behind their attitudes may include inadequate availability, labeling and marketing of environmentally beneficial products, as well as higher prices for some of them'.

This again demonstrates the variability between studies according to the regression model used, or if such a model is used and the strength of environmental attitudes in predicting behavioural change. Minton and Rose (1997) show this to be the case perfectly with their example of green buying in the southern United States. When personal norms were controlled for, environmental attitude correlated more significantly with behavioural intention, whilst personal norms acted directly upon behaviour itself. Thus the direct effect of environmental attitude was weaker, but its indirect effect was still strong.

Environmental Attitude and Value Scales: An Evaluation

The evidence given above lends credence to the idea that environmental values do have an impact on both behavioural intention and behaviour itself when seen in the context of the psychometric scales presented above.

However, the extent to which these scales are reliable and the variability in the correlative values between attitude and behaviour is problematic. This difficulty may be associated with scale validity or genuine disparity between environmental values and actions (particularly if high scores (i.e. high skewness) is demonstrated on such scales). One difficulty that has become apparent is the extent to which scales vary according to the population sampled and the impact this might have upon an analysis. Tarrant and Cordell (1997) analysed the differences in the attitude-behaviour relationship for five well-known environmental attitude scales in Tennessee, United States. They found significant differences between the attitude-behaviour correlation according to gender (generally females having higher correlations), education (higher educational level resulted in higher correlations) and income (lower income earners tended to have higher attitude-behaviour correlations). Although the authors state that use of scales should consequently be adjusted according to the population sampled, this implies that the scale is at fault and not the fact that gender, for example, might be a more important predictor of environmental behaviour than global attitudes. This is emphasised by Wall (1995) who studied Canadian attitudes towards local pollution issues. She found that education and political affiliation were both linked strongly to general environmental concern. Although she found general widespread concern for the environment, these population differences were still important (a case put as early as 1981 by Van Liere and Dunlap). For further examples of this phenomenon, the reader is referred to a number of articles that have investigated the underlying structure of environmental concern (Arcury *et al.*, 1986; Bord and O'Conner, 1997; Buttel and Flinn, 1978; Dietz, *et al.*, 1998; Dunlap and Mertig, 1995; Gamba and Oskamp, 1994; Jones and Dunlap, 1992; Klineberg *et al.*, 1998; Krause, 1993; Norris, 1997; Samdahl and Robertson, 1989; Stern *et al.*, 1993; Van Liere and Dunlap, 1980). Although this is partially relevant to this discussion, it should be noted that interest here lies principally with the elucidation of variables that explain behaviour and so for purposes of space, discussion of this literature is omitted.

It therefore appears that two issues emerge from this discussion. First, environmental values scales cannot be assumed to be unidimensional; they are complex and must be treated with care whilst under construction and after fieldwork is complete. Second, they cannot be seen in isolation as dominant predictors of specific environmental behaviours, but must be seen in the context of a host of other variables that might both impact on their formation and modification with regard to both behavioural intention to act *and* behaviour itself.

Environmental Values in the United Kingdom

As Skrentny (1993) has shown, environmental concern in the UK lags far behind that of other European nations. Asked the question 'Would you like to see more or less government spending for the environment?' only 37% of Britons stated that they wanted much more or more to be spent, as opposed to 83% of Germans and 73% of Austrians requesting this amount of spending. Even more Americans (44%) said they wanted much more or more spending on the environment. This situation reflects what is well known among the academic and wider community in Britain: the UK lags behind Europe on a number of crucial social welfare and environmental issues, especially where the individual citizen (or in the British case, subject) is concerned. Such low levels of concern could be put down to an already healthy sum of money being spent of the environment, but as has been the case for some time, the UK spends comparatively small amounts of money on the environment compared other nations. Thus far this chapter has detailed North American studies that have shown support for environmental attitude and values scales, although as has been noted, there is less certainty about the relationship of these scales to environmental behaviour, as discussed above. However, a number of studies have been undertaken in the UK to elucidate the 'level' of environmental concern amongst the people of Great Britain. These have focused around the annual publication of the *British Social Attitudes* series (other research exists, but not in the same psychometric format as the work from elsewhere). These have sometimes yielded different results from those of Skrentny (1993) alluded to above (although it should be remembered that his was a cross-national study).

Witherspoon and Martin's (1992) study of British environmental attitudes showed that three principal concern factors emerged, namely: the global environment, pollution and nuclear power. Scores were high on all scales and were skewed heavily towards categorising the problems as 'very serious'. However, as they note, this may be more to do with giving the 'environmentally-correct' answer (a difficulty highlighted by Scott and Wilits, 1994, above). Nevertheless, their research did appear to show severe disquiet with the state of the environment. This is also the case with Norris' (1997) analysis of the 1990 British Social Survey. He defines three principal types of environmentally concerned person ('old-green', 'new-green' and 'anti-nuclear'). However, he acknowledges that the split in the data means that the broad idea of 'Environmental Concern' is less applicable since different specific interests are involved.

This was also evident in a different way with Bryan Taylor's (1997) analysis of questions dealing with solutions to environmental problems. Attitudes declined sharply when specific measures to curb environmental problems were stated. Hence concern is seen at the most general of levels, but less at specific levels. This disparity between acceptance of long term general environmental aspirations and opposition to immediate 'stick' measures to improve environmental quality was demonstrated in Christie and Jarvis's (1999) analysis of attitudes towards three fundamental environmental issues – countryside development, transport use and access to rural land. They concluded that there was an enduring support for the principles of environmental protection in these three instances and favourable acceptance of 'carrot' measures to implement them (such as improved public transport). However, support for 'stick' measures was severely lacking (such as congestion charging in cities), demonstrating perfectly the extent to which people are likely to endorse measures that have a regressive impact on their perceived quality of life (such as being able to drive their car freely).

This is partly confirmed by Dalton and Rohrschneider's (1998) analysis of the International Social Science Programme database. Not only did Britain score low on the NEP as compared to other European nations, but it also conformed more to the traits of materialism, outlined by Inglehart (1977, 1981, 1990). These facts, along with the high correlation between the NEP and materialism scores, led the authors to conclude that environmental concern was linked more to social values (materialism) than any objective assessment of the environment. This finding would explain the results achieved by Witherspoon and Martin (1992), Norris (1997) and Christie and Jarvis (1999) given above, since it might be argued that whilst general concern does not conflict with material values, more specific attitudes on the environment may conflict with materialist values. Nevertheless, the work of Witherspoon and Martin (1992) still shows a higher mark up of values than the Dalton and Rohrschneider (1998) study.

Clearly environmental values are diverse and complex in the UK. It would appear that as more specificity in environmental attitude scaling is attained, concern decreases. However, as Witherspoon and Martin (1992) point out, there is a risk that people are giving what they perceive to be the correct answer. Nevertheless, given the preceding discussion concerning environmental values and behaviour, it might be expected that if there were a link between environmental values and action, this would reflect these scores on attitude scales. Certainly environmental behaviour is much lower in the UK than other developed nations (Dalton and Rohrschneider, 1998), but the extent to which this can be attributed to orally given attitudes about the environment is uncertain.

Cross-national Aspects of Environmental Values

Attention should be drawn to the potential differences between values, in particular environmental values, at the cross-national level. As was demonstrated above, environmental concern and scores on certain value scales has been shown to be significantly lower in the UK than in mainland Europe and the United States. Unpacking the reasons for this apparent fundamental difference in priorities and values is not simple. The issue may come down to one of specificity, as was the case with some of the British research outlined above. It was found that as specificity increased, concern decreased. Given that most British studies have used tests based on concern (i.e. how concerned are you that x is a problem?), it may well be that if the same types of questions were asked of Americans or Europeans with the same regularity the results would be different. The lower score on the NEP found in Dalton and Rohrschneider's (1998) study of British social attitudes may be the result of what Barr (1998) remarked was the essentially American nature of the test which as a result may not have been culturally appropriate for the UK.

All of these deliberations point to a fundamental problem when attempting to compare environmental values cross-nationally. The differences in measurement instruments renders any informed hypothesis regarding value-based differences obsolete. However, it is worth providing a brief commentary regarding the possible differences between North American, European and British environmental values. To begin with North America, there is certainly some argument for splitting along Canadian and United States values. The development of what Selman (1994) has termed the Canadian 'environmental citizen', to which we shall return below, implies the internalisation of environmental values which arguably have not been demonstrated within the United States. The US is at both extremes of environmentalism. The current President is reviled the world over for his rejection of the Kyoto Protocol on climate change and his attempts to drill for oil in a protected Alaskan national park. Representing what is essentially the Texan view of the environment, President Bush is juxtaposed to the clean-living and organically-minded liberal thinking residents of California. Yet, this raises somewhat of a contradiction, since Los Angeles is one of the most polluted conurbations on the planet.

The argument then comes full circle, in that although some Americans must aspire to clean living and a sustainable lifestyle, the extent to which this is undertaken and championed by a reduction in car use is questionable. North and Central Europe may present a closer relationship between aspiration and reality. The Green Party has a strong hand in a number of European nations as well as the European Parliament. In Britain, the poor showing of the Green Party, the marginalisation of environmentalists as 'freaks' and 'weirdoes' provides the backdrop for poor recycling rates, high car use and a low ranking for the environment in political terms.

Does this necessarily mean, however, that Germans and the Swiss are intrinsically more green than the Texans or British? As yet this question cannot be answered satisfactorily, but what the discussion above shows is that environmental values and behaviours are measured by so many different standards, it is impossible to state definitively either way. What is clear, however, is that in most nations, aspirations very rarely reflect reality and that whilst environmental values logically underlie behaviour, they are merely the starting point when examining environmental action.

3 The Social Psychological Perspective II: structural and situational variables

Introduction

The second set of factors that are related to environmental behaviour, outlined above in Chapter One, concerns the variables that are essentially the objective circumstances of the individual. An excellent review of these factors has been given by Schultz *et al.* (1995) and the reader is referred to this paper for a brief inventory of the data presented below. The variables have been used in differing ways and with varying frequency to express the impact of non-personal characteristics upon environmental behaviour. Four key categories can be used to encompass the diverse factors involved:

- *Context*: The extent to which geographical and/or contextual factors affect behaviour;
- *Socio-demographics*: The extent to which distinct populations behave differently;
- *Knowledge*: The extent to which global and specific knowledge impacts upon action; and
- *Experience*: The extent to which behavioural experience affects current activity.

In addition, the reader is referred to what is now becoming somewhat of a dated article, but one that is often quoted as the authority on determinants of environmental behaviour – the meta-analysis by Hines *et al.* (1987).

Context and Environmental Behaviour

In the arena of waste management, 'context' essentially concerns two aspects of recycling behaviour that have largely been ignored in the literature. First, there is often an issue concerning the provision of a kerbside recycling facility to residents. Second, there is the issue of how far residents must travel in order to use facilities such as bottle banks. In essence, the former is a question of either being with or without a given service, whereas the latter is a function of both actual and perceived distance to recycle wastes. The second of these concepts (perceived distance) is dealt with in Chapter Four below. However, here the discussion is focused on the objective structural aspects of recycling schemes. There have been very few studies into this geographical phenomenon and this is therefore an under-researched aspect of environmental behaviour that requires more attention.

Two good examples of the importance of the distance and provision of recycling receptacles are given by studies by Ball and Lawson (1990) and Barr (1998) in the UK. Barr (1998) examined seven spatial areas in the Vale of White Horse District of Southern Oxfordshire in order to assess the degree to which recycling provision affected recycling behaviour. He found that there was an almost linear relationship between level of recycling provision and recycling behaviour. Those in the most remote areas with the lowest level of provision had recycling scores roughly one-third of those in areas with the best level of provision. Ball and Lawson's study of recycling behaviour in Scotland examined nine areas according to the degree of urbanisation (city, urban or rural) and levels of recycling performance. As Barr (1998) has outlined in Oxfordshire, rural areas had fewer recycling services and so the rural-urban continuum formed a useful provision measure for Ball and Lawson. Consistent with the findings of Barr, rural dwellers made journeys to recycling sites less often, with a reduced range of materials. The relationship was further supported by the fact that rural residents appeared to have similar scores on other items, including publicity campaigns. Hence, the distance, effort and logistics involved were obviously important in shaping rural dwellers behaviour.

Berger (1997) has examined the degree to which kerbside collection can impact on recycling behaviour in Canada. Berger found that although socio-demographic criteria initially provided the best explanation for the data, when kerbside collection was added as a predictor, these relationships faded away. This demonstrates that the relationships found in numerous previous studies, which failed to look at spatial elements, may have been registering spurious relationships.

The work of Derksen and Gartell (1993) and Guagnano *et al.* (1995) has addressed this theme more fully and has examined the impact of kerbside recycling in the context of other linking variables. Derksen and Gartell (1993), in their study of recycling in Calgary and Edmonton, Canada, found that access to a kerbside recycling facility was the dominant predictor of recycling. Indeed, those who did not appear to be concerned with the environment recycled when access to a blue box was given. However, even those who were concerned about the environment did not recycle until they had access to a blue box. Once this opportunity had been fulfilled, then recycling levels were higher than those with a blue box, but with lower levels of concern.

This is in contrast to the work of Guagnano *et al.* (1995) in their study of Fairfax County residents in Virginia State. There, attitudes had very little impact upon recycling for those with kerbside collections, but a large effect for those without. This apparent contrast could be down to study design differences, as acknowledged by the authors. However, it raises the question of the extent to which attitudes (on the general environmental level) have an impact when recycling is not readily available. Nevertheless, these studies demonstrate the importance of examining spatial and contextual differentiation in recycling provision as a crucial predictor of recycling behaviour.

Socio-demographics and Environmental Behaviour

A further structural set of variables that have been found to impact upon environmental behaviour and waste management in particular are the personal circumstances and traits of the individuals involved in such behaviours. These have been found to centre around the following variables:

- Age;
- Gender;
- Education;
- Income;
- Political affiliation; and
- Other variables.

Age

Conventional wisdom has held that those involved in environmental behaviours are young people (Hines *et al.*, 1987). The justification for this argument is that young people are more highly educated about the importance of environmental activity and are more politically liberal – another correlate of pro-environmental behaviour. However, as Schultz *et al.* (1995) point out this is not necessarily the case.

A younger population involved in environmental behaviour has been reported by a number of researchers. Weigel's (1977) study of pro-ecology behaviours (a set of five including participation in a recycling programme) in New England demonstrated the correlative value of age. He found a negative relationship between an index of the five behaviours and age, asserting that younger people were more likely to be involved in the behaviours (and also noted a negative relationship to age and liberalism, as noted above). Sia *et al.* (1985) analysed an environmental behaviour inventory (including waste behaviour) in Illinois and demonstrated the difference in age between members of the Sierra Club and Elderhostel members. Using the former as a proxy for positive environmental action and the latter as a proxy for lack of action (Sia *et al.* term them 'samples of conveniences'), they found that the average age range of the Sierra Club was 35-45 years, whereas the average of the Elderhostel was 55-65 years.

This was again supported by Oskamp *et al.* (1991) in their analysis of recycling behaviour in California. However, unlike the preceding studies discussed, they only found a negative predictive relationship between recycling for a monetary incentive. When it came to voluntary kerbside recycling or other environmentally responsible behaviour, there was no predictive value either way for age. Hence, it appears, as Oskamp, *et al.* (1991, p. 313) point out, that:

> ...different environmentally responsible behaviors have different patterns of antecedents.

This is a theme that will be referred to later in Chapters Six to Nine. However, once again this demonstrates the value of the methodology utilised by Oskamp *et al.* The previously cited studies have merely recorded a correlation or descriptive difference between age and behaviour. Because they have been unable to classify the importance of that relationship they have been unable to assess the importance of age against all other variables. Given this fact, interpretation of studies that have not classified behaviours should be used with caution.

In contrast, a number of studies have alluded to a positive relationship between age and pro-environmental behaviour. Schahn and Holzer's (1990) analysis of individuals in Heidelberg, Germany demonstrated that there was a weak positive correlation between age and scores on their 'self-reported actual commitment' index (including seven environmental topic areas, such as resource conservation). Vining and Ebreo's (1990) study of 197 households in Illinois also demonstrated that those who they termed 'recyclers' (defined as those who stated they had recycled 'something' in the past year) had higher average ages to those who had not. In a similar way, Lasana (1992, 1993) noted that those who maintained recycling over a given number of months in a pilot recycling project in New York State tended to be between 40 and 64 years old, the non-recyclers being below 40. In a similar model used by Oskamp *et al.* (1991), Derksen and Gartell (1993) also found a positive relationship between the number of items recycled by residents in Alberta, Canada. Again, however, the emphasis placed on the predictive power of this variable is far less than the unclassified bivariate studies listed above. In this case age had only a weak (if significant) positive effect upon recycling behaviour. Baldassare and Katz's (1992) study of four environmental behaviours (including recycling at home) found a positive relationship between age and reducing driving behaviour, green buying and saving water (recycling at home was positive, but not significant). However, in the same vein as the previous authors, they found that the predictive value of age was overshadowed by non-demographic factors.

In the UK, Ball and Lawson's (1990) study of recycling behaviour in Scotland noted that younger people were those least likely to participate in recycling programmes, whilst McDonald and Ball (1998) state that in Glasgow a disproportionate number of elderly people tend to recycle plastics.

Finally, age has been found, on occasions, to have no predictive or correlative value whatsoever. Tucker's (1978) study of general environmental behaviour, Steel's (1996) analysis of environmental activism and Daneshvary *et al.*'s (1998) study of textile recycling all indicated no relationship at all. Gamba and Oskamp's (1994) study of recycling in Claremont, California, reported a negative bivariate relationship between age and self-reported recycling. However, when this was added into the hierarchical regression its significance was nil. This shows the value of this regression analysis technique (Oskamp *et al.*, 1991).

Of course, these are only studies that bother to report the relationship between age and behaviour. Nevertheless, it appears that the results are mixed. However, given the range of analytical techniques involved it is hard to see the wood for the trees. Oskamp *et al.* (1991) have summarised the situation well, as quoted above, in stating that different environmental behaviours are likely to have varying predictors. Hence, age is likely to correlate to some behaviours positively and some negatively. Indeed, as Oskamp *et al.* showed, 'recycling' or 'waste management' behaviour cannot be seen as just one type of action, but a number.

Gender

Considerable research exists concerning the difference between men and women regarding environmental values and the reader is referred to the list of references on this subject given in Chapter Two above. However, interest here rests with the relationship of gender with environmental behaviour. Very few studies report the statistical difference between gender and environmental behaviour. Schahn and Holzer (1990) did find a positive correlation between gender and active commitment to conservation activities, implying that women were more likely to partake in such activity. This was also found by Witherspoon and Martin (1992) who studied the *British Social Attitudes* survey and found that men were significantly lower scorers on their 'Consumer behaviour' scale, asking respondents about their consumer habits. This indicated that women were more likely to buy environmentally friendly products consciously. Gender also became important in Baldassare and Katz's (1992) study of four environmental behaviours in California, where women were more likely to (and were important in predicting) reduce driving behaviour, buying environmentally safe products and conserving water (although not recycling at home). This was also found in Steel's (1996) analysis of environmental activism, where females were significantly associated with predicting behaviours such as petition-signing and campaigning on environmental issues.

However, surprisingly Blocker and Eckberg's (1997) analysis of the United States 1993 *General Social Survey* failed to find any relationship between gender and environmentalism. They note that:

> Contrary to the expectations of gender theory, women (and men) of higher social status, with more knowledge, and with greater trust in science are more likely to engage in proenvironment action and are less likely to see the economy as more important than the environment. We find only extremely weak direct effects of gender ...on the propensity to take part in personal pro-environment actions like recycling (Blocker and Eckberg, 1997, p. 854).

Their assertion is supported by Daneshvary *et al.* (1998) who found no relationship between gender and textile recycling. This is also shown in Gamba and Oskamp's (1994) analysis of recycling in California.

As Van Liere and Dunlap (1980) have shown with reference to environmental concern, although the correlations tend to be reasonably weak, there are consistently positive results showing that women are more likely to be more environmentally concerned than men. It is therefore surprising that this does not appear in the behavioural research. However, since such little reporting of the impact of gender has been undertaken and since this reporting has been principally of a bivariate nature, the jury remains out on the effect of gender on environmental behaviour.

Education

Many more studies have analysed the impact of an individual's formal education upon environmental behaviour. These mostly show positive correlations, implying that higher educational level equates with enhanced behaviour. Wiegel's (1977) previously cited study of pro-ecology behaviour found a very positive correlation between public education and his index of behaviour. Similarly, Sia *et al.* (1985) were able to report that members of the Sierra Club in Illinois were all educated to at least college degree level, whereas those in the (less-environmentally active) Elderhostel club were only educated to basic college level. Schahn and Holzer's (1990) group of environmentalists also had higher levels of education, although the coefficient was low. Berger (1997) found a similar weak, but positive, correlation with recycling paper to educational level within his study in Canada. Indirectly, but nonetheless importantly, Vining and Ebreo (1990) found higher education important for guiding the sources of knowledge used by recyclers to gain information, which thus had a knock-on effect to environmental behaviour.

Regression models, used to assess the predictive value of a given variable when others are taken into account, include Lasana's (1992) study of New York recyclers, where education, along with age and income, was a significant predictor of being a recycler. Derksen and Gartell (1993) found education a significant, albeit minor, predictor of recycling scores in Alberta, Canada. Baldassare and Katz (1992) showed a similarly weak, but positive relationship between education and recycling in California. This was also the case with Steel's (1996) study of environmental activism, where education was a positive, but not significant, predictor of environmental action. However, Oskamp *et al.* (1991) again showed the value of categorising environmental behaviours. Education had a significant predictive value for kerbside recycling, but not for other environmentally responsible behaviour or recycling for cash. Hence, although the majority of studies appear to show that environmental behaviour and in particular recycling, can be predicted (quite logically) by incorporating education as a factor, this may only be relevant in certain circumstances. Indeed, as Gamba and Oskamp (1994) and Daneshvary *et al.* (1998) have demonstrated, education has no impact whatsoever on textile recycling in Nevada nor commingled recycling in Claremont, California, respectively. Hence, it appears that education does have a general overall positive impact upon recycling behaviour, but that this relationship is more subtle than most previous research has been able to show.

Income

Again, income has been widely positively associated with high levels of environmental concern, although this does not necessarily reflect across the globe (Dunlap and Metig, 1995). Evidence to support the income-behaviour hypothesis comes from Berger (1997), who reports a low to moderate correlation between the two (although it should be remembered that Berger reported modification effects of recycling provision). Gamba and Oskamp (1994) do not have the same qualifying point attached to their study of household recycling in Claremont, California. Using hierarchical multiple regression, they found that income of residents was a significant, if rather weak, predictor of observed recycling. Watts and Probert (1999) also made a link between higher income status and recycling behaviour in Swansea, South Wales. Using three areas of the city defined primarily by wealth status (i.e. household income levels), they found evidence that the principal determinant (indirect) of recycling was residential area. They link this directly to their socio-demographic proxy for affluence. However, as following sections will demonstrate, other factors might be involved in shaping what might be (here, at least) a spurious relationship. In another study from the UK, Ball and Lawson (1990) found a similar relationship in their analysis of nine residential areas' glass recycling habits in Scotland. They found that users of the static recycling schemes tended to be from higher socio-economic groupings than non-users.

McDonald and Ball (1998) however, found that with regard to plastics recycling in Glasgow, a high proportion of participants (nearly 50%) came from economically inactive groups. Well under a third of those in the highest economic categories took part. This appears to counter the two previous articles' findings, although as has been noted before, different levels of behavioural specificity and/or different behaviours are likely to have different antecedents and therefore any concrete conclusion is hard to make.

Nevertheless, Baldassare and Katz (1992) found evidence that income was significantly related as a predictor variable to reducing car use, although not, it should be noted, to recycling or their other two environmental behaviours. Daneshvary *et al.* (1998) also found that income was negatively related to textile recycling in Nevada. However, other American examples of the income-behaviour relationship are not convincing. Vining and Ebreo (1990), Lasana (1992) and Derksen and Gartell (1993) all showed no relationship whatsoever between income and recycling behaviour. Oskamp *et al.* (1991), using their hierarchical regression model, demonstrated that whilst income appeared to have a relationship with kerbside recycling, this was not borne out by the regression results.

It therefore appears that although income might have the perception of being a predictor of recycling and other environmental behaviour, it is variable, probably (although this cannot be proven here) because of the differences in other factors and the behaviours being examined.

Political Affiliation

Again there has been less work concerning the effect of political affiliation regarding actual behaviour as opposed to environmental concern. However, Riley Dunlap's (1975) classic paper is a milestone in the confirmation of what had been suspected in and outside the United States during the environmental crisis of the 1970s. Students from the University of Oregon were asked to rate their environmental action according to three scales: action on one environmental issue, action on a local environmental issue and support for environmental rights. As was expected, consistently higher behaviour scores for more liberal individuals were rated. This was the case for Democrats against Republicans and Liberals against Conservatives. Such a result was replicated by Weigel's (1977) previously referenced study of pro-ecology behaviour in New England. Both Liberal and Socialist respondents correlated to environmental behaviour with 0.34 and 0.42 Pearson coefficients respectively, whilst conservatives correlated with a − 0.48 coefficient.

Schahn and Holzer's (1990) German study later confirmed both of these results as still holding true, where 'left' political values correlated with conservation behaviour. Indeed, Daneshvary *et al.* (1998) report a similar situation in Nevada eight years later. In slight contrast, Baldassare and Katz (1992) found that the effect of political affiliation was mixed according to the environmental behaviour being examined. Recycling had a slightly (although insignificant) relationship with conservatism, whilst reducing water use had a significantly positive relationship. Yet a negative relationship was found when reducing car use and buying green were examined.

Attention has focused more on examining environmental values (e.g. the BSA studies). Witherspoon and Martin (1992) however examined the issue of political affiliation with regard to their 'Green consumer' scale and found that those voting either for the Green Party or (to a lesser extent) the Liberal Democrats, were more likely to buy environmentally friendly products and buy with an environmental conscience. Indeed, a reading of the popular press leaves one in no doubt where the most vocal of activists place their political allegiances. O'Riordan (1997a, 1997b) has discussed the poor performance of the Labour government with regard to environmental policy, but in the 1980s the Labour Party certainly was the home of environmental activism for the mainstream in Britain, especially anti-nuclear enthusiasts (O'Riordan, 1985). However, since the sliding of environmental issues from the political agenda post-1992, the effect of party political affiliation on environmental behaviour is less simple to judge. Nevertheless, given the strong links between liberalism and environmental action demonstrated above, it might not be too tentative to suggest that most seriously concerned environmentalists vote either Liberal Democrat or Green (the former having a good record on environmental protection in local government).

Other Variables

A number of other variables have been mentioned only briefly in the literature concerning both environmental concern and behaviour. However, these will be dealt with briefly here.

First, in terms of *socio-demographic* factors, there is the issue of household composition and tenure. Oskamp *et al*. (1991), Lasana (1992), Berger (1997) and Daneshvary *et al*. (1998) all find this a significant variable. Oskamp *et al*. (1991) report that this positively predicts both kerbside recycling and other environmental behaviour. Lasana (1992) found that those who recycled more also tended to be those who owned their own home. Finally, this was also a positive predictor for Berger (1997) and Daneshvary *et al*. (1998). This was also found to be an issue in Swansea by Watts and Probert (1999), where those who privately owned their own home were more likely to participate in kerbside recycling schemes. However, Gamba and Oskamp (1994) who measured this variable found no evidence to support the ownership hypothesis.

Type of dwelling also has an impact on recycling behaviour, as found by Berger (1997), who noted that those in apartments recycled less than those in single family dwellings (an important predictor for Derksen and Gartell, 1993, also). Indeed, although mentioned very little, there is the negative behavioural effect of lack of car access on waste behaviour (Barr, 1998).

There is also the issue of occupational status and environmental behaviour. This is more of a methodological than philosophical point and concerns the extent to which basic occupational strata have an impact upon environmental behaviour reporting. Many of the studies so far reported use college students and it is arguable that this group is by no means representative of the population as a whole. Hence, since this and other occupational groups (e.g. the retired) often form quite identifiable groups in society, any variation in behaviour among these groups might be useful to analyse, since policies aimed at changing their behaviour would simpler to implement. Of course, the lack of studies including such basic occupational data is a problem, but it would provide a worthwhile exercise since such groups might reflect indirect effects previously unseen.

Second, a small number of papers have discussed the impact of '*religiosity*' on environmental behaviour. Wiegel (1977) found a strong relationship between those who showed traits of religiosity and pro-ecology behaviour, whilst Guth *et al.* (1995) demonstrated a similar relationship. A difference should be stressed between religiosity and formalised Christianity or Islam. As Campbell (1999) has stated, the growth in spiritual and mystic movements is clearly linked to what he terms the 'Easternisation' of Western belief systems, connected to 'environmentalism' and 'consciousness-raising'. The drop in those believing in 'a personal God' has been compensated for by the proportion of those who believe in 'some sort of spirit or life force'. This is what is meant by 'religiosity' and Campbell's (1999) thesis implies that the 'Easternisation' of Western beliefs systems has led to a rise in the intrinsic link between religious beliefs and concern for the environment. Conversely, as noted by Young (1990), institutional religions, such as Judaism and Christianity, have advocated the view that humans have dominion over nature. This led Young (1990, p. 62) to note that:

> It was a short step from the emphasis on the permission of God to utilise the environment to human purposes to the belief that technological advancement

and success in moulding the environment to human purposes was the reward of Protestant virtue or evidence of God's approval.

Thus it is too simplistic to argue that 'religion' either influences environmental behaviour in a positive or negative manner, but rather that there may be a more subtle positive link between spiritual religiosity and behaviour and thus a somewhat negative link between Judeo-Christian institutional beliefs and behaviour. Such a categorisation is evidently too basic, but it is important to note that there is some evidence for a 'religion-behaviour' link.

Conclusion

In essence, socio-demographic variables evidently do have an impact on environmental behaviour, in particular recycling behaviour. However, these vary according to the behaviours (range and detail) being studied. Indeed, as Berger (1997) has pointed out, researchers must be careful when assigning predictive value to such variables, since their actual relationship could be quite spurious. Hence, there is a need to be more specific about the behaviours being researched, as well as using appropriate techniques to assess the predictive power of such variables in the context of other factors that could impact upon environmental behaviour (Oskamp *et al.*, 1991).

Knowledge and Environmental Behaviour

Information relevant to the behaviour in question is without doubt a prerequisite for undertaking that behaviour. However, when correlations of knowledge to behaviour are not strong, this does not necessarily imply a weak knowledge-behaviour relationship. Rather there could be issues of questionnaire design or specificity that cause a somewhat unclear relationship. Finding the difference between these two states of affairs is not simple. However, as might be expected, behaviour-specific knowledge has been found to be a crucial predictor of environmental behaviour in previous research. More abstract knowledge (see below) may be less relevant, although this is in question (Schahn and Holzer, 1990; Hines *et al.*, 1987).

Ever since Weigel and Amsterdam's (1976) demonstration of the impact of behaviour-relevant knowledge in producing changes in attitude-behaviour correlation coefficients from 0.11 to 0.55, researchers have tried to measure behaviour-relevant knowledge without having to resort to experimental designs. Sia *et al.* (1985) found a significant role for the impact of knowledge in their analysis of environmental behaviour (including waste behaviour) in Illinois. Using stepwise regression techniques they found that knowledge of specific environmental behaviours was a significant predictor of their behavioural index. However, although a significant predictor, knowledge fell far behind perceived skill and environmental sensitivity as a predictor of behaviour.

Kallgren and Wood (1986) examined the impact of access to behaviour-relevant knowledge in their study of attitude-behaviour consistency. They found that those with high levels of behaviour-relevant information had more stable attitudes and behaviours and that consistency was high within this group. They also surmised that given the more behaviour-relevant information, respondents were motivated to undertake recycling activity because of the environmental benefits, whereas those with lower levels of accessibility to this information perceived recycling for more aesthetic reasons (e.g. street cleanliness).

Simmons and Widmar (1989-1990) also demonstrated the importance of behaviour-relevant knowledge in their study of mandatory recycling in New Jersey State. Although they found high participation amongst certain (popular/well-known) items, such as newspaper and glass, many did not recycle more new and peripheral (behaviourally speaking) items. For example, 49.9% did not recycle motor oil. They put this down to the fact that:

> It is reasonable to expect, however, that people are failing to adopt these practices because they lack sufficient knowledge and understanding of how to incorporate them into their daily lives (Simmons and Widmar, 1989-1990, p. 328).

Vining and Ebreo (1990) supported this view in their analysis of recyclers and non-recyclers in Illinois. They identified a number of traits that are important when considering the effect of knowledge upon recycling behaviour. They argued that in their study recyclers were better informed, had more knowledge of the availability of local recycling programmes and gave more accurate information regarding materials that could be recycled. Non-recyclers, in contrast were unsure and vague about recycling. Vining and Ebreo (1990, p. 68) offered a tentative explanation for this phenomenon by stating that:

...nonrecyclers selectively ignore or discount information they perceive as being irrelevant to their own behavior, whereas recyclers seek out and remember information about recycling.

Nevertheless, it appears that behaviour-relevant knowledge is crucial to guiding recycling behaviour. Indeed, as Vining and Ebreo (1990) and Kallgren and Wood (1986) have pointed out, access to this information is also crucial and Vining and Ebreo also noted that recyclers had a larger number of sources of information than non-recyclers. Lasana (1992) found this in her study of New York State recyclers and non-recyclers. Using discriminant analysis she found that awareness was a key means of discriminating between the two behavioural groups and that recyclers tended to use newsletters about recycling and newspapers as sources of information about recycling.

Finally, Gamba and Oskamp's (1994) study of recycling behaviour in Claremont, California, showed the importance of knowledge in promoting commingled recycling programmes. Knowledge was by far the largest predictor within the hierarchical regression model of observed recycling. Gamba and Oskamp (1994) asserted that this finding was not surprising since attitudinal factors were unlikely to have an impact in an activity in which so many people were involved. Hence, it appears that behaviour-relevant knowledge plays a larger part when recycling is more of an established behaviour, in this case at least. Indeed, Oskamp *et al.* (1991) have acknowledged the importance of knowledge in their study of the non-commingled scheme in Los Angeles. However, it was far less prevalent in predicting kerbside recycling behaviour in that case than in Claremont.

Thus far the discussion has focused upon behaviour-specific knowledge, but Schahn and Holzer (1990) have emphasised the need to distinguish between what they term 'Abstract' and 'Concrete' knowledge. The former concerns the basis of environmental knowledge, for example underlying knowledge regarding the state of the environment. The latter constitutes what Schahn and Holzer term 'knowledge for action'. In other words, behaviour-specific knowledge required to act appropriately. They found that in their study of environmental behaviour in Germany (including waste behaviour), variation in abstract knowledge had no bearing on variation in self-reported behaviour, yet concrete knowledge had a large effect. They argue that the apparent lack of consistency between knowledge and behaviour reported by some authors (a good example being Maloney and Ward, 1973) is that the measurement used was too abstract as compared to the behaviour being measured. Nevertheless, although this study did not find any direct relationship between abstract knowledge and behaviour, there is certainly evidence that such a relationship (direct or indirect) might exist (Hines *et al.*, 1987). It seems quite logical to expect that higher levels of general environmental knowledge will have more of an impact on environmental behaviour.

Experience and Environmental Behaviour

This remaining structural variable has had relatively little attention paid to it (Fazio and Zanna, 1978). However, a number of studies have found that experience of pro-environmental behaviour in the same arena can have a significant positive impact on another behaviour. Macey and Brown's (1983) analysis of energy saving behaviour demonstrated the value of using past habitual experience as a predictor of current behaviour. They examined three energy saving behaviours with reference to previous experience with those environmental behaviours and concluded that:

> The more familiar the individual is with the behaviour through direct past experience, the more likely that individual is to engage in the behaviour again (Macey and Brown, 1983, p. 138).

Luyben and Bailey's (1979) and Goldenhar and Connell's (1992-1993) analysis of paper recycling also drew this conclusion. However, as Macey and Brown point out, there is a problem if the behaviour has not been undertaken before. Hence, whilst being academically interesting, for those not already involved in the behaviour, it is not relevant when attempting to encourage new behaviours.

Nevertheless, workers in this area have attempted to examine the relationship between experience in one environmental behaviour and action in another (Kallgren and Wood, 1986). Oskamp *et al.* (1991) have examined what they termed 'behavioural factors' (i.e. behavioural experience) within their model. They found that energy conservation factors and other environmentally responsible behaviours predicted recycling for cash. Experience did, however, negatively predict kerbside recycling behaviour, a fact that might reflect the proportions of those with and without experience in the sample, rather than a negative relationship. Similarly, Daneshvary *et al.* (1998) found that textile recycling was related to the extent to which participants already undertook other waste recycling behaviours. Werner and Makela's (1998) longitudinal study reported that those who had higher levels of recycling experience, or 'recycling history' were correlated well to both initial levels of other recycling and recycling at later sampling intervals. This indicated that:

> ...respondents with strong personal and social attitudes and positive phenomenal experiences were most likely to describe multiple ways of organizing recycling and to report fewer interferences to recycling (Werner and Makela, 1998, p. 382).

Although very little research exists incorporating experience as a variable, it is logical that this would have some impact. Of course this is likely to vary according to other factors. However, as a basic premise, it seems likely that action in one environmental behaviour, if that is voluntary, may lead to indirect or direct effects on other behaviours. The processes under which such influence might be exerted are not alluded to here, but it might be anticipated that the combination of enhanced understanding of an activity would improve an individual's psychological perception and acceptance of an activity sufficiently to realise the consequences of acting differently in other behavioural realms.

Structural Variables: Conclusion

Clearly there are a large number of variables that have been correlated to and used to predict environmental behaviour. The more complex models have permitted a more thorough examination of these variables, which has meant that in many cases apparently concrete correlations have been proven spurious. Of course there are still problems in terms of the difference in study areas, populations, instruments, representativeness and so on. However, what is certain is that there exists a need to examine these situational variables. It is crucial that further research into this area provides two important sets of information. First, it must ensure that the relationships identified are not spurious, but real. Second, wider reporting of socio-demographic results would ensure a more valuable insight into research articles.

It is also important to accurately define what is meant by 'situational' or 'structural' characteristics since at the beginning of the chapter they were defined as those factors that on the most part were objective situational circumstances of individuals. This evidently is not strictly true. Individuals can actively seek out more information and of course change demographically. However, in grouping these four variables (context, demographics, knowledge and experience) together, a definitive set of factors has been created that define personal circumstance based on objective criteria at a given point in time, rather than subjective perceptional or psychological criteria, the focus of which is the subject of the next chapter.

4 The Social Psychological Perspective III: psychological factors

Introduction

The third set of predictor variables identified concerns a group of factors that may be termed 'psychological' variables. A large body of research exists that has sought to predict environmental behaviour according to a number of social-psychological theories and individual personality variables. These are grouped together here for clarity. Among the theories that have been used to predict waste-related behaviour is Schwartz's (1977) model of normative influences on altruism. It has been used widely to provide models of recycling and environmental behaviour. Individual variables include the use of subjective, the role of personal and response efficacy, the importance of environmental threat and the influence of logistical problems on behaviour.

The term 'psychological' is used here to express those factors that are specifically individual in nature, such as perceptions or personality traits. For example, it might be argued that those with more altruistic personalities would be more likely to be involved in environmental behaviour since they would have a wider concern for those people and objects around them with less emphasis placed on the benefits or disadvantages from acting. This personality trait can be contrasted to perceptional data which are interpretations of a given state of affairs by an individual. Perceiving, for example, that individual actions have little efficacious impact is a perception; it is an interpretation of the effect of a behaviour which may be arrived at by drawing on a series of informational and communicative sources, such as radio, TV and newspaper reports, individual conversation and previous experience. Both types of factors are significant. They will affect how an individual views a given environmental action in both the way it is seen generally and the consequences of acting in particular.

Such factors are quite different from those previously investigated in the chapters above. They comprise possibly the most diverse and complex set of variables which mean that those wishing to actively change environmental behaviour face a significant task in unpacking personality and perceptional characteristics in such a way as to attempt to manipulate action without being accused of control freakery. As such, the term 'psychological' is a broad one, but encapsulates the necessary meaning.

Schwartz's Model of Normative Influences on Altruism

Schwartz's (1977) classic theory of the normative influences on altruism has been used by a number of authors to investigate moralistic elements of environmental behaviour. Before examining these studies in detail, an outline of the theoretical model of Schwartz will be given. It should be noted that Schwartz outlined his model as a broad social-psychological framework and not as a purely environmentally based model. What he describes as a 'processual model' (Figure 4.1) of activation of moral obligation to altruistic behaviour begins with an initial 'Awareness of need'. If the individual is unable to assess that there is a problem or dilemma in a given situation, then there will be no understanding of need on the part of (in Schwartz's case, the other individual) a given subject. If this requirement is met, then the model shows that the next state of consciousness that the individual requires is the 'Perception that there are actions that could relieve the need'. It is plausible in many human situations that although a great need of another individual or object is perceived, the recognition by others that there are actions that could provide relief is lacking. Assuming that these two preliminary conditions are met, Schwartz envisages a further step, namely 'Recognition of being able to provide relief'. Logically, although a person may perceive a need and recognise a way of alleviating that need, they may not register a way in which they can personally act. Only if this condition is met can a person move to an 'Apprehension of responsibility' to become involved, by which the individual personalises the situation such that responsibility is located in the self. As can be seen in Figure 4.1, this stage interacts with the personal and situational norms of the individual to activate feelings of moral obligation. It is at this point that the individual begins the stages of serious assessment of the situation, beginning with the 'Assessment of costs'. If these are positive (or more importantly, not negative in any way), then the individual moves to perform the alleviating behaviour. However, should there be questions at this stage, the individual reassesses the situation, with possible 'Iterations of earlier steps'.

The rather brief description of the theoretical model of Schwartz given above is only a basis for the work to be outlined below, but the framework has been used extensively by social scientists to determine whether a large number of behaviours, including environmental action, are altruistic and influenced by moral norms (Thogersen, 1996). However, the majority of studies that have applied the Schwartz model have usually used far fewer constructs than in the schematic model outlined in Figure 4.1.

One good example is the work of Van Liere and Dunlap (1978). They assessed 'Awareness of need' by what they call 'Awareness of consequences' when analysing yard burning behaviour. In evaluating the ascription of responsibility to oneself they used 'Apprehension of responsibility'. Both constructs were measured by statements, followed by a 'Do you agree/disagree?' option. For example, 'Awareness of need' (or in their case, 'Awareness of consequences') was measured by asking:

> Some people say that smoke from backyard burning makes it difficult for people with respiratory problems to breathe. Do you agree?

Figure 4.1   **Schematic   representation   of   Schwartz's   (1977) 'processual' model of normative influences on altruism**

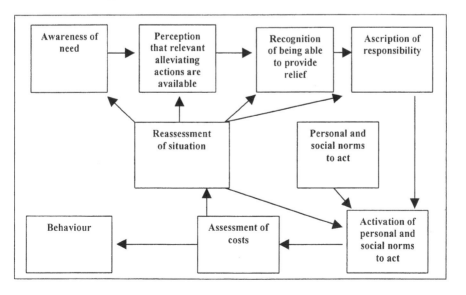

Quite obviously, the question prompts the respondent in assessing whether there is any 'need' on behalf of people who suffer as a result of yard burning. They found that those individuals with high levels of 'Awareness of consequences' and an active sense of responsibility to act were statistically more likely to not use yard burning. However, as with the Schwartz model, they found that those who felt a high responsibility to act were much more likely to reduce yard-burning behaviour. The issue of environmental responsibility is dealt with below as a separate predictor variable.

Guagnano *et al.* (1994) used three constructs of the Schwartz model to assess the Willingness to Pay (WTP) for public goods, based on a 'contribution' model, where individuals put money 'in' to pay for something communally, rather than paying individually to take something (selfishly) 'out'. The authors hypothesised that this model of contributions would be explained more fully by the altruism model, since contributing may be seen as a moralistic behaviour, requiring the activation of personal and situational moral norms. As with Van Liere and Dunlap (1978) 'Awareness of need' and 'Apprehension of responsibility' were gauged. However, the authors also assessed what they termed 'Personal social costs', which correlated to 'Assessment of costs and outcomes' in Figure 4.1. All three constructs were measured by using a battery of attitude statements. They found that, as with Dunlap and Van Liere, awareness of need raised willingness to pay, but that this was modified by awareness of responsibility and crucially personal costs. Hence, this additional variable was also significant in predicting a willingness to act.

Guagnano *et al.* (1995) used a more complex method to assess their causal model of recycling behaviour. The constructs of the Schwartz model were tested by means of a ten-item, four-point response format, assessing various factors. The causal framework did not include any notion of 'Awareness of need' to recycle, but rather focused on the perceived consequences of recycling, the personal costs of that behaviour and the ascription of responsibility. It is important to note that in the model developed by these authors, the ascription of responsibility comes *after* assessments of consequences of the behaviour, whereas this variable comes *before* such assessments in the framework of Schwartz (1977). However, given that the study focused on the prompting of behaviour (recycling bins), we might expect that there would be some change in the model structure, differing as it does from the idea of Schwartz that his model be used as a measure of individual 'helping' behaviour toward another individual. Guagnano *et al.* (1995) found that the constructs in the Schwartz model worked as predicted for those without access to a kerbside recycling bin, but that this effect was significantly reduced after recycling bins were introduced. This implies an interaction between the structural and psychological variables so far not mentioned.

Perhaps the most well known study in this area is the research into recycling behaviour in Denver by Hopper and Nielsen (1991). Using an experimental recycling programme as their base, they measured the fundamental aspects of the Schwartz model. They concluded, as with Van Liere and Dunlap (1978), that environmental behaviour was governed by an awareness of consequences of action and the personal responsibility involved.

The Schwartz model provides a social-psychological framework very different from other models of behaviour (see Chapter Five below) (Thogersen, 1996). However, although it is acknowledged here that awareness of consequences and acceptance of responsibility to act is important, the degree to which such a framework can offer an holistic model is limited. As Blamey (1998) has demonstrated, there are other variables that can be added to the Schwartz model that enhances its predictive power. Stern *et al.* (1985-1986) also question the validity of the model in the context of different moral actors. Indeed, there is also the difficulty that additional structural variables could fundamentally change the efficacy of a predictive model. Nevertheless, it is important to acknowledge the mechanics of the Schwartz model.

Intrinsic Motivation, Satisfaction and Environmental Behaviour

A further psychological aspect of environmental behaviour to be examined here is the extent to which people undertake conservation activities because they receive a degree of satisfaction from undertaking that behaviour. Raymond De Young is the major proponent of this thesis. The central theme in De Young's model is that:

> 'People seem to derive considerable satisfaction from the very activities that others try so hard to encourage them to do' such that '...conservation can also be perceived as contributing to one's sense of satisfaction' (De Young, 1986, p. 447).

In many of De Young's papers, outlined below, he has tried to assert that there is no real difference between 'recyclers' and 'non-recyclers' in terms of their views towards recycling, extrinsic motivation or the effort that recycling entails, but rather that intrinsic motivational factors, such as being frugal, are more prevalent in distinguishing between 'conservers' and 'non-conservers'.

To support his thesis, De Young has undertaken a number of studies of individual attitudes and behaviours towards energy saving and recycling behaviour. In one of his first papers on this subject, De Young and Kaplan (1985-1986) conducted thirty open-ended interviews with residents in a small city in the United States. The interview asked questions about everyday activities, which De Young and Kaplan then used to interpret the structure of satisfactions derived from energy saving behaviour. They argued that those who tended to conserve energy the most were those who could find intrinsic motives to do so and more importantly, to continue to do so. They concluded that:

> This suggests that conservation behavior might be found potentially satisfying to a broader cross-section of the population (De Young and Kaplan, 1985-1986, p. 240).

In a more statistical vein, De Young's (1986) later paper dealt with intrinsic motivations (satisfactions) of recycling. He outlined eighteen satisfaction items that he saw as important to recycling, including avoiding waste, repairing, self-sufficiency, participation, etc. Again, De Young defined a certain number of key broad satisfactions within his sample: frugality, luxuries and participation. Interestingly, his data showed that the apparently contradictory variables of frugality and luxuries were not necessarily opposed and that:

> ...environmentally appropriate behaviour may be made to appeal to a broad cross-section of Americans rather than just to people of a Spartan nature (De Young, 1986, p.447).

Again, De Young concluded that derivation of some satisfaction from recycling, whether this be feeling good about conserving resources or making a community contribution by participating, seemed to be important in motivating and maintaining recycling.

In a similar piece of research, De Young (1985-1986) undertook another mail questionnaire, asking respondents a total of thirty-nine satisfaction questions and fifteen items on motivations for recycling. The satisfaction items were grouped into frugality, participation and prosperity, whilst the motivational items were grouped as non-materialism, intrinsic motivation and extrinsic motivation. Again, De Young found that satisfaction and intrinsic motivation played a large part in shaping personal behaviours. Support for these findings and the notion that there are no large value-orientated differences between recyclers and non-recyclers was provided by a later study, where De Young (1988-1989) undertook a telephone survey, comparing two groups of individuals by means of similar satisfaction scales as used in his previous work. Finally, De Young's (1990) review of various studies into recycling behaviour in Michigan also provided support for his thesis. The studies asked questions regarding the motivations to recycle, including conservation of natural resources, helping a charity, earning money and 'because it's the right thing to do'. Once again, De Young found that non-monetary, intrinsically-motivated satisfactions were the ones most regularly outlined by people as being crucial in their motivation to recycle.

Wider support for De Young's thesis is not easy to find, since few researchers have used 'satisfaction' as a correlative or predictor variable in their research. Oskamp *et al.* (1991) used an aggregated scale that tapped elements of intrinsic motivation, which in hierarchical regression analysis was shown to be the third most significant predictor of kerbside recycling behaviour and was also a significant predictor of other recycling and environmentally responsible behaviour. Gamba and Oskamp's (1994) study of commingled recycling in California demonstrated that an environmental concern factor, including items such as 'satisfaction in saving natural resources' was a significant predictor of reported commingled kerbside recycling behaviour. Mckenzie-Mohr *et al.* (1995) found that there was a positive statistically significant difference between those who gained feelings of satisfaction from home composting and those who did not and the frequency with which it was undertaken. Finally, Werner and Makela's (1998) longitudinal study of recycling behaviour showed that satisfaction and feeling good about recycling was crucial in maintaining what they termed 'trivial' behaviour. In this sense, such a variable can be seen to be important in making a laborious activity interesting and bearable.

It is highly likely that those who derive intrinsic motives for recycling (i.e. because they enjoy it and it makes them 'feel good') are more likely to recycle in voluntary static schemes than those who appear to derive extrinsic satisfaction (i.e. monetary incentives/legal penalties) for recycling or not recycling, respectively.

Subjective Norms and Environmental Behaviour

Subjective norms refer to what Fishbein and Ajzen (1975) regarded as a primary influence on behaviour. The awareness of others' behaviour, along with an acceptance of that behaviour, was, according to Fishbein and Ajzen, a crucial factor affecting personal decision to take action. We might see such a construct as a form of social influence that affects all human beings at regular intervals. Tucker (1999b) has outlined a model of normative influences on kerbside recycling behaviour using receptacle set out data from kerbside collection schemes in Scotland and the North West of England. By comparing set out rate data and subsequent frequency and weight of kerbside recycling over time, Tucker formulated an empirical model that enabled him to predict set out rates of households. He acknowledged that this method could not prove that set out rates/weights influenced others to adjust their behaviour. Nevertheless, his data pointed to a significant predictive value of the model using frequency of set out as its major predictor.

Other studies have been more conventional in their assessment of the impact of social norms in predicting recycling behaviour. Werner and Makela (1998) showed that in their longitudinal study of recycling behaviour, awareness of neighbours recycling and actual reported neighbours recycling were both highly correlated to recycling at their first time of sampling, although this influence dropped later in their study. Oskamp *et al.* (1991) asked respondents whether their friends and neighbours recycled material in their evaluation of kerbside recycling in California. Using hierarchical multiple regression, they found that social awareness of others' recycling was a significant predictor of kerbside recycling, as well as being less significant in predicting other environmental behaviour. As Tucker (1999b) has found, Oskamp *et al.* (1991) have reported that overall the behaviour of significant others around individuals enhanced kerbside recycling almost more than any other variable measured.

Similarly, Gamba and Oskamp (1994) found what they termed 'pressure', referring to social influences, had a significant predictor of commingled recycling. Nevertheless, this was much reduced in importance compared to knowledge of the recycling programme, concern for the environment and logistical issues. Indeed, Vining and Ebreo (1990) found that 'social reasons' for recycling were not distinguishable between those who did and did not recycle.

Despite this somewhat surprising conclusion, there is evidence from the experimental literature reviewed below that social influence and pressure can have positive effects on pro-environmental behaviour. Studies into 'Block Leader' programmes, where significant members of the community lead efforts to enhance recycling, have been found to influence recycling behaviour (Everett and Pierce, 1991-1992; Burn, 1991), as well as introducing normative feedback interventions (Schultz, 1998).

Social norms do play a part in the modification of behaviour. However, this appears to vary (as it would be expected to) according to the research context and the degree of predictive value in the framework being used.

Environmental Threat and Environmental Behaviour

Environmental threat is cited by a small number of authors as being significant in predicting behaviour towards the environment (Baldassare and Katz, 1992; Steel, 1996; Segun *et al.*, 1998). Baldassare and Katz (1992, p. 604) assert that:

> '...environmental threats overshadow youth, high education, high income and liberalism as predictors of overall and specific environmental practices'. They also state that '...personal threat of environmental problems is predicted by the demographic and political factors that have, in the past, been linked to other measures of environmental concern' (p. 605).

Inasmuch, their assertions are worthy of examination, since it may be that environmental threat provides a mediating influence to the more structural factors involved in predicting environmental behaviour. The data used by the authors include a large telephone survey, in which environmental threat is measured by means of a single question regarding threats to the individual of environmental problems. Using four environmental behaviours as the dependent variables (including recycling waste), Baldassare and Katz found that for their sample of Orange County residents in California were more likely to take part in all four environmental practices if they perceived that issues such as air quality and water pollution could have a detrimental impact upon their health and well-being. This was more important than demographic or political factors. As found by Berger (1997), the apparent relationship between demographics and behaviour was in fact not predictive, but spurious. In the case of Baldassare and Katz, young, female individuals and those with a liberal political leaning were most likely to feel threatened by the environmental threats given above.

Apart from this one study that put environmental threat at its centre, there are no other studies yet known that place this variable at the heart of environmental attitude-behaviour research. Segun *et al.* (1998) used scales to assess the threat posed by environmental problems to health and related this to environmental activism. The types of scale used were a list of specific environmental problems that could be seen as threatening to health, such as air quality and bacteria in food. Their path analytic model demonstrated that perception of the health risks involved was directly related to environmental action taken.

Nevertheless, there have been few examples of the perception of 'threat' being correlated to environmental behaviour, probably because it has been assumed that individuals have internalised the impacts upon the global environment much more readily than they have concerning the impact of an environmental problem on themself. However, it has been shown clearly by Baldassare and Katz (1992) that if there is a perception that well being or health are threatened, then this can act to override conventional predictors of environmental action. It is perhaps difficult to conceptualise how unsustainable waste management might be perceived as, for example, a threat to well being. However, if this is taken to mean quality of life, then it is more plausible that people might be more actively concerned with waste issues by the impending construction of a new incinerator. Hence, environmental threat may provide an innovative means by which to investigate the degree to which individuals have personalised statements of concern about waste (i.e. have they acknowledged waste as a problem to themselves as well as a national/global issue of concern?) Of course with such a gap in the literature this cannot be proven one way or the other here, but it does provide evidence that in certain situations there might be a role for personal threat playing a part in shaping pro-environmental behaviour.

Response Efficacy and Environmental Behaviour

As described in Section 4.1 above, Ajzen and Madden (1986), Kantola, *et al.* (1992) and Chan (1998) have demonstrated that personal (or 'self') efficacy is a predictor of both behavioural intention and actual action within the context of the Fishbein-Ajzen model of behaviour. However, a number of other studies have examined the importance of a variation on this factor. There should be a distinction drawn here between self-efficacy (referring to personal control over events or actions) and response-efficacy (referring to the perception of how a given behaviour will impact to reduce a given problem).

Arbutnot (1977) has examined response efficacy in his study of recycling behaviour in Ohio, United States. As part of a large set of personality items, Arbutnot measured the respondents' belief that their actions were of little or no worth since the magnitude of the environmental problem was too large and that environmental degradation was a problem that should be tackled by large corporations. He found that those involved in recycling were statistically more likely to believe that their actions had tangible results. Again, as mentioned above, there are problems with this type of comparison, since the predictive power of the variables were unknown. Nevertheless, Arbutnot argued that such differences were greater and more significant than differences between the recyclers and non-recyclers concerning pro-ecology attitudes, indicating some power of the variable to distinguish the two groups.

Becker *et al.* (1981) studied residential energy consumption in the United States in the late 1970s and undertook a questionnaire of 207 couples in households. Using objectively measured energy consumption as their dependent variable, they determined that identifying one's individual role in the energy crisis, in particular believing that personal actions made a difference, was important in explaining variation in both summer and winter energy use. Nonetheless, it should be noted that the explanation offered by their regression model was very poor, accounting for just 18%.

In contrast to this study, Oskamp *et al.* (1991) found no evidence to support the notion that what they term 'efficacy re: environmental problems' had any effect directly upon recycling behaviour. However, they did find that this variable had a high predictive power for variables that had a predictive power for behaviour, such as intrinsic motives to recycle.

Hence, those studies that have examined response efficacy found that this is an important variable, although this may range in significance and directness in its prediction of behaviour. It is logical that the more a person feels able to make a difference environmentally, the more they are likely to take action.

Logistical Factors: Time, Convenience, Storage Space and Behaviour

A number of studies have researched the significance of logistical factors that the individual identifies as real (or perceptional) barriers to individual environmental action. The three principal factors are time (perceived time to do a given action, in comparison to other behaviours), convenience (the perceived simplicity of undertaking a given behaviour) and storage space (a factor unique to waste management dealing with the issue of storing recyclable material).

Vining and Ebreo's (1990) analysis of recyclers and non-recyclers distinguished between these two groups heavily according to what they termed 'nuisance' factors, which they gave as distance to drop-off site, no kerbside collection and lack of time. These factors along with the need for financial incentives were the most important in distinguishing the groups attitudinally. Gamba and Oskamp (1994) also found this in their recycling study. Correlation analysis revealed significant negative correlations between their personal inconvenience factor (defined as no space for storing recycables, the attraction of pests by having materials for recycling in the house, no time to prepare recyclables, difficulty in defining what is recyclable and issues of moving the bin to the kerb) and behaviour. Hierarchical regression analysis revealed that this was reduced in importance as compared to knowledge and environmental concern. However, it was obviously a significant factor.

Lasana's (1992, 1993) research into recycling behaviour also revealed that inconvenience was a major predictor of problems associated with the recycling scheme studied. The most significant perceived problem was the lack of a kerbside collection scheme, a major component of inconvenience. However, respondents also found general inconvenience and the costs involved in recycling a major problem. This was also the case in a study of composting behaviour by McKenzie-Mohr *et al.* (1995). They found that composters varied significantly from non-composters according to the extent to which they felt composting was unpleasant, inconvenient and time consuming. A wide ranging study of why Chicago residents did not take part in recycling was carried out by Howenstine (1993). Using factor analysis, he found that three principal factors governed the reasons why residents did not recycle: nuisance factors (mess, inconvenience and space to store recyclables), spatial factors (location of static site) and indifference. Tucker (1999a) also reported that issues of time, effort and storage were important in his study of recycling in Scotland.

Again, it would be surprising if issues of time, convenience and so on were not issues for certain members of the population, although this is not always the reason given for non-participation (Ball and Lawson, 1990). These are, of course, *perceptions* of time, etc. and not objective measures of inconvenience and so on. However, it is apparent that some people do feel these issues are important and this may be related to other factors, like lack of information or relevant knowledge. It appears that whatever the underlying reasons, logistical considerations are important to evaluate when examining environmental behaviour.

Environmental Citizenship and Appropriate Ecological Behaviour

This factor has been neglected in the attitude literature and has been touched on in studies that have sought to examine environmental responsibility as a significant factor in predicting environmental behaviour (e.g. Arbutnot, 1977; Oskamp *et al.*, 1991). This brief review will introduce the subject of citizenship before considering environmental citizenship as a variable.

Defining Citizenship

Bryan Turner (1993, p. 2), an eminent scholar on the theme of citizenship, has defined citizenship as:

> ...that set of practices (judicial, political, economic and cultural) which define a person as a competent member of society, and which as a consequence shape the flow of resources to persons and social groups.

Within this framework, Turner discussed the notion of the city and nation state as the institutions within which citizenship can take place. Anyone lying outside these institutions becomes automatically excluded and loses a key 'stake' in society. Turner acknowledged that this notion of citizenship was innately ethnocentric and that there is an alternative to the 'traditional' concept of citizenship, that of the 'plural' citizen. In other words, citizens could be different, thus disabling the nation state from being the 'handle' of citizenship. Already, therefore, a number of key problems that are faced by the student of citizenship have been identified. However, these are debates that must be acknowledged, but not dwelt upon. The more pressing task here is to discuss recent trends in Anglo-American thoughts on this issue, encapsulated by the work of Amatai Etzioni (1993, 1995a, 1995b). Well known as a philosophical mentor of Tony Blair, Etzioni's (1993, p. x) 'Communitarian' agenda has sought to address the '...cold season of excessive individualism...'. This is illustrated by the:

> Increasing rates of crime, illegitimacy, drug abuse, children who kill and show no remorse...and political corruption... (p. x).

Using what is essentially a moral argument, Etzioni (1993) has called for a moral revival, but without all of the throwbacks of traditional puritarianism. Aspects of this crusade include: law enforcement, but avoiding a police state; protection of the family, without banishing women to the home; ensuring moral education in schools; allowing people to live in 'communities' without fear; ensuring that individuals equate strong responsibilities with ever-increasing amounts of rights; and balancing self-interest with duty to the community. Again, it is not the remit of this chapter to pick the Communitarian thesis to pieces, but rather to give the reader a coherent overview. In that case, there is only one salient point that needs to be made in the context of the current chapter and that is the importance of rights and responsibilities. Etzioni (1993) argued that there must be a change from:

> ...a demand that the community provide more services and strongly hold up rights - coupled with a rather weak sense of obligation to the local and national community (p. 3).

Later, when writing on moral elements of Communitariansism, Etzioni raised the issue of community in the context of rights:

> Our society is suffering from a severe case of deficient we-ness and the values that only communities can uphold; restoring communities and their moral voice is what our current conditions require (p. 26).

Balancing the rights of an individual within a community, with the responsibilities that the individual has to that community, is the subject of part of Etzioni's later work (1995b). Taking de Tocqueville as his guide, Etzioni (1995b, p. 22) suggested that:

...the best protection against totalitarianism is a pluralistic society laced with communities and voluntary organisations, rather than a society of highly individualised rights carriers.

So what of these communities? As Etzioni acknowledges (1995, p. 24) his definition of 'community' has on some occasions been rather 'fuzzy'. However, he explains the concept in terms of nests, where each community is placed within a larger one. Individuals can be in more than one community at any one time and most crucially 'Community need not be geographically concentrated' (p. 24).

Smith (1999) recently provided a coherent account of the place of geography in such debates, attempting to 'deconstruct' communitarianism. He identified two strands of communitarianism: conservative and liberal-democratic. The former is criticised on the grounds that all that is required by society is a return to the pre-modern lifestyles that encapsulates the 'myths' of romantic, homogenous communities. The latter suffers from the fact that although there is some 'difference' allowed within communities, the essence of such communities is that they are closed to 'non-members' and as such there is a distinct possibility of conflict between communities. Etzioni (1995a, 1995b) has certainly overcome some of these problems, but the fact remains that such theses will always tread the tightrope between liberalism and conservatism. Indeed, communitarianism is more often than not accused of the former, encompassed in the cries of a return to the 'Nanny State' by opposition politicians in the United Kingdom. Smith also focuses on some of the community-based assumptions of Etzioni's thesis and notes that although the all-encompassing notion of community sounds appropriate, there may be conflict between different geographic and non-geographic elements of community (e.g. the residential community and the workplace community, respectively). This point was emphasised in an environmental context by Agyeman and Evans (1995) who stated that although citizens may feel a sense of community, there is no reason to believe that there will not be conflict in the way that people wish to represent and administer that 'place'. Indeed, a note of caution was also struck by Painter and Philo (1995) concerning the cultural aspects of citizenship. Although Etzioni has vehemently denied a (re)turn to majoritarianism, these authors questioned the extent to which 'others' like gays, lesbians, the young, old, black, etc. would be able to take a full part in the Communitarian crusade. Painter and Philo noted that citizenship in the modern world should mean, above all, that anyone can walk freely in any public space and feel comfortable.

From this summary, a number of points can be made. First, citizenship, in the true meaning of the word, must involve rights with responsibilities. Second, some sense of community, from which and to which such obligations are held must be present. Third, citizenship should mean inclusivity of all citizens. Now that the basics of citizenship have been established, there is a need to look more closely at these ideas in the context of the environment.

Environmental Citizenship

Newby (1996) has given an historical account of the growth in environmental concern in the UK. Although he identifies four periods of 'concern', it is only the last that concerns this discussion. The era of sustainability is upon us and what could be more indicative of such an era than the term "our common future" (WCED, 1987)? As Newby has asserted, the emphasis focuses on a common and global agenda. However, as he also noted, there are not the institutions to cope with the demands of such an agenda. If we are all global citizens, in terms of the environment at least, what are our rights, responsibilities? Who in our communities is included and excluded?

Rights and responsibilities   are dealt with comprehensively in a global context by Waks (1996) who utilised T. H. Marshall's theory of citizenship and adds a fourth layer of 'environmental' rights to the list compiled by Marshall (civil, political and social). He has discussed the different types of rights that are available concerning the environment. First, there are basic human rights of the environment that are prerequisites for any human being living in a community. Second, there is a right to environment, which states that people have certain environmental rights in parallel to other human rights, such as the right to clean air. Finally, there are the deep ecology notions of rights *of* the environment, which assert that the environment itself has rights.

In terms of responsibilities, a closer look at the local level is necessary to find work that has been done in this area. Selman (1994, p. 46) has discussed the concept in terms of:

...the duties associated with active citizenship to include responsibilities for the environment.

He emphasised the need for people to realise that living in the global community brings certain behavioural responsibilities. However, this would not be composed from an authority above, but aided by the inclusion of an informed citizenry in the decision-making progress. Only then will people feel responsible. What Selman and Parker (1997) have termed 'active' citizenship, involving responsibilities as well as rights, is distinctly lacking in the UK today, partly to do with cynicism in government circles and partly understood by public individualism. Having said this, Macnaghten and Jacobs (1997) did find some evidence for public feelings of responsibility, but this is watered down by a lack of personal agency (a factor of importance in the work of Eden, 1993). As can be seen from the above, the rights we are due as environmental citizens and the obligations we have as citizens, are very unclear, although this is not surprising since the concept is a relatively new one.

Communities and participation   are seen as key elements of the process of moral decision making that will involve the environment. However, as Selman (1996, 1998) and Selman and Parker (1997) have acknowledged, the sense of community in the UK is distinctly lacking at present. If LA21 is really going to work, participation of the whole community working as a community is required. Macnaghten and Urry (1998) noted the same phenomenon, arguing that people tend to live in the 'here and now', rather than plan for the future. The feeling that a sense of community has been lost was certainly present in their focused interviews and this affected personal attitudes toward action and the degree to which any action would assist the environment. Selman (1996, 1998) and Selman and Parker (1997) have noted an important key trend in the development of LA21. There is not so much the *en*rolment of citizens in environmental activities, but rather *re*enrolment. Community champions, committed greens and local authority officers are all drafted in to help, but they have all done it before (Selman 1998). In other words, LA21 is simply another elitist bandwagon upon which the good and great of the community can leap. Can this really be called 'citizenship' if only certain people take part? Petts (1995) highlighted these problems in a study of community decision making in Hampshire. The formation of 'Citizen Panels' in the study had a number of disadvantages. First, they were biased toward very environmentally aware citizens, disenfranchising those who were not as aware. Second, the belief among those who did take part was that they were only really PR exercises by the local authority. The nature of the fora led to little time for decision making and discussion. All in all, they were ineffective in achieving one of the main aims of LA21 – participation (see also Murphree *et al.*, 1996, for an American example). However, a different scheme involving the National Forest in Leicestershire had more positive results, leading Bell and Evans (1997) to remark that:

> ...the forest initiative reinforces established links in Britain between the aesthetic appreciation of, and training about, nature and the country's 'enterprising' past, as prerequisites for 'responsible' citizenship (p. 270).

However, such 'national' schemes are rare and engendering similar amounts of enthusiasm into local projects may be more difficult.

In essence therefore, there are two key variables that appear to be of importance when discussing the modification of environmental behaviour from the citizenship angle. First, a right to a clean and safe environment must be balanced with tangible responsibilities at the individual level for that environment. Second, the conditions under which such responsible environmental behaviour should occur are within networks of communities that have adequate representative processes whereby individuals feel they have an active role in decision making on the environment.

As has been mentioned above, the literature from which the citizenship thesis emanates is primarily qualitative and no attempt has yet been made to integrate the elements of the citizenship thesis within the framework of a quantitative research strategy. However, it seems logical that those who treasure their environmental rights and feel personally responsible for the environment are more likely to act positively, especially when elements of community and democracy are most favourable.

Psychological Variables and Environmental Behaviour: Conclusion

There are a diverse number of studies that have examined personality variables and environmental behaviour. These range from models of environmental action, like the adapted Schwartz model, to the linking of individual variables to environmental action. There is overlap between the models on the one hand and the individual variables on the other. For example, Schwartz's (1977) framework postulates that 'Acceptance of responsibility' is placed within a causal framework that has both antecedents and further requirements before this is translated into behaviour. Arbutnot (1977) has meanwhile merely correlated responsibility with behaviour. Indeed, responsibility can also be seen in the context of environmental citizenship and behaviour (Selman, 1996). Hence, differing bodies of literature have conceptualised similar variables in differing methodological and theoretical moulds to suit their own needs. Nevertheless, as has been shown, they all have merit, but they lack a structured framework that forms an umbrella to encompass the diversity shown.

Clearly this group of variables has importance when examining environmental behaviour. They are distinct from the other groups in that they play on the purely subjective elements of environmental behaviour, focusing on the extent to which an individual engages with a number of attitudes towards the action, such as a belief in the efficacy of the behaviour and the social acceptance of the action. These are characterised by personality traits, such as altruism and the various elements of, for example, intrinsic motivation, which affect the way people view the means to achieve the goal of behaviour. Alternatively, perceptional elements, such as response efficacy and environmental threat, impact on the way in which people view their response to the behaviour.

Having now completed this analysis of the three fundamental groups of factors that influence environmental behaviour, it is necessary to examine how these seemingly independent groupings can be organised to formulate a framework of behaviour.

5 Social Psychological Models of Behaviour: conceptualising action

The Psychological Model

The previous three chapters have displayed the range of variables that researchers have tested in relation to environmental behaviour. Many of these researchers are pure psychologists; others have used psychological methods to examine the efficacy of certain variables on behaviour. The dichotomy between these two approaches poses somewhat of a problem for the student of environmental behaviour. Without doubt the rigorous training within the discipline of psychology provides the most appropriate background for examining the determinants of human behaviour. Yet other social researchers have identified factors, as demonstrated previously, that cannot be ignored when providing an account of environmental action. What splits these two alternative approaches fundamentally, however, is the use by some psychologists of 'models' of human behaviour and the use by other social researchers of 'frameworks' of behaviour. The difference is subtle and yet crucial in terms of the underlying assumptions that are made about how human beings operate.

The psychological 'model' is essentially a way of explaining action according to a set of rigorously defined criteria that have been theoretically justified and posit a well-defined outcome given a combination of variables within the model. Examples of such models include the well-known Theory of Reasoned Action (TRA), Theory of Planned Behaviour (TPB), Schwartz's model of normative influences on altruism, as well as other specifically defined models for use in particular studies. Such models have a limited number of input variables and pre-defined relationships between such variables that posit a certain outcome that in turn will determine the efficacy of the model. For example, the model of altruistic behaviour given in Figure 4.1 posits a behavioural outcome on the basis of a process beginning at 'Awareness of need' and ending at 'Behaviour'. The model is closed and there is minimum flexibility. This is inherent in other such models because the authors are inevitably trying to demonstrate the efficacy of a given theory of behaviour that they have developed.

These psychological models can be contrasted by alternative 'frameworks' of behaviour. This loose term can be used as an umbrella label to incorporate most other research into environmental behaviour. The crucial difference being defined here is the flexibility of a framework over a model. Frameworks are conceptualisations of behaviour based less on theoretical understanding and more on 'what works'. They may often be informed by theoretical research, but more often than not are defined by their 'ad hoc' nature. Examples of such frameworks include the work of Oskamp *et al.* (1991) and Gamba and Oskamp (1994) who examined the impact of various sets of factors on kerbside recycling behaviour which, although not limited by a processual logic, were defined by a hierarchy of factors that informed their analyses. However, although this hierarchy was implicit, a finding to the contrary would not mean that their frameworks were a failure, since there was not a linear process under examination. Indeed, the recognition by social researchers that 'other factors' might be important is probably the defining feature of those who utilise the 'framework' approach.

Both of these approaches have their merits and disadvantages. The psychological model benefits from a clear logic that can be proved or disproved easily. If proven, such models can provide a unique understanding of human behaviour that can be immensely useful in informing policy makers about how to potentially change behaviour. However, this is perhaps where the problem with the psychological model becomes most acute. The social researcher must ask him or herself whether it is correct to predict human behaviour in such a way that essentially restricts the human potential for irrationality. The assumption that human action can be predicted, like weather patterns or floods, is fundamentally flawed and is a notion that social researchers should keep at arms length. This is not to state that advocates of psychological modelling necessarily argue that this can be the case, but rather that users of such models misinterpret the apparent simplicity of the model in question.

Such a conclusion would imply that it is fruitless to examine the diversity of human behaviour and its potential causes, since human action is not totally predictable. Yet this postmodern argument pens an even longer suicide note than the modelling argument. As discussed in Chapter One, the notion that humans are unique entities that must be understood as individuals on all levels is a compelling, logical and seemingly watertight argument. However, although the notion of individualist decision making provides a logical end point to the debate, it does not provide a solution to some of the world's most pressing environmental problems. Hence, whilst the modelling of human behaviour may seem overly determinist on the one hand, the postmodern advocates of individualism are defeatist on the other.

The framework approach may offer some middle ground between determinism and defeatism. The benefits of using flexible frameworks to examine the underlying determinants of behaviour are that far fewer assumptions are made concerning the process of behaviour creation and change and that the addition or omission of factors is acceptable in different situations. Although data may be placed in a hypothetical hierarchy, such a hierarchy can be dismissed if found to be inappropriate. Indeed, this can be achieved without having to effectively throw the baby out with the bath water. The framework approach also offers a means by which to examine and identify underlying trends in behaviour that cannot be undertaken if there is the assumption that all individuals are different. Hence, the fundamental tenets of the approach offer flexibility and loose assumptions, but on the basis of rigorously collected and analysed data.

At the same time, this approach can be criticised on two alternative fronts. From the position of a psychological modeller, the approach can be seen as weak and to offer nothing in terms of theoretical progress on understanding human behaviour and consequently predicting it. From the postmodern standpoint, the approach still provides a largely determinist model of behaviour that is flawed because it assumes that generalisations can be drawn across human populations.

It is argued here that the framework approach, despite the limitations presented above, provides the social researcher with the most reliable means of examining human behaviour. What the approach loses in predictive power and individualist understanding, it gains in flexibility and generalisation. The following section describes how such a framework can be developed from a theoretical basis and how this can be applied to formulate a framework of waste management behaviour.

Conceptualising Environmental Behaviour in the United Kingdom

The preceding discussion concerning psychological models of behaviour and frameworks for behavioural analysis provides a basis on which to construct a conceptualisation of waste management behaviour in the United Kingdom. As has been noted in previous chapters, the vast majority of research into environmental behaviour within the UK has been within the context of qualitative and, more often than not, postmodern interpretations of behaviour. It is the contention of this book that such research is of great value when considering the underlying meaning of, for example, sustainable development and environmentalism. However, when attempting to understand what drives a particular behaviour, it is necessary to examine more quantitative techniques that can yield reliable data of a standardised nature over a large sample population.

Within the social psychological genre of literature that has been the subject of Chapters Two to Four, the two main approaches described earlier in this chapter have been utilised. Psychological models, such as the TRA or Schwartz's normative model have been used extensively (Olsen, 1981; Kantola *et al.*, 1982; Jones, 1990; Hopper and Nielsen, 1991; Guagnano *et al.*, 1995, to name a few). Frameworks have been used to examine the relative effects of one or maybe two of the three sets of factors identified in the preceding three chapters upon behaviour.

There has as yet, according to the author's knowledge, been no research that has examined the effects of all three sets of factors on behaviour within a framework that has been based also on some theoretical understanding of action. It is argued here that if a framework developed with some recourse to an existing model of behaviour, but not restrained by that model, could be developed, then such a conceptualisation would be liable to testing as with a conventional model and theory, but not be seen as a failure if other factors were deemed important.

This challenge is difficult to grasp, but it can be argued that one psychological model in particular offers a unique understanding of behaviour that can be significantly adapted and loosened to accommodate the requirements of a flexible framework that can also be used to incorporate all three sets of factors examined in Chapters Two to Four. The Theory of Reasoned Action, mentioned above, provides just such a psychological model and the sections below describe firstly its nature and structure and then proceed to assess its efficacy in explaining behaviour.

The Theory of Reasoned Action

Fishbein (1967) and Fishbein and Ajzen (1975) have developed the *Theory of Reasoned Action* (TRA) which has been widely acclaimed (and latterly dismissed) by social scientists as a means of predicting a given behaviour from a set of hierarchically linked constructs (Figure 2.1).

Working from right to left, in reverse order to the model itself, 'behaviour' is the action in question that one seeks to predict. 'Behavioural intention' is the next step back and this is used to predict behaviour directly. This is simply a measure of how an individual intends to act towards the object in question. Hence, Fishbein and Ajzen argue that, for example, intending to vote Democrat in a given election should predict actually voting for that party. However, although this link is virtually without question in the model, the predictors of behavioural intention are twofold and likely to vary in importance.

Figure 5.1   **The *Theory of Reasoned Action* (Source: Fishbein and Ajzen, 1975)**

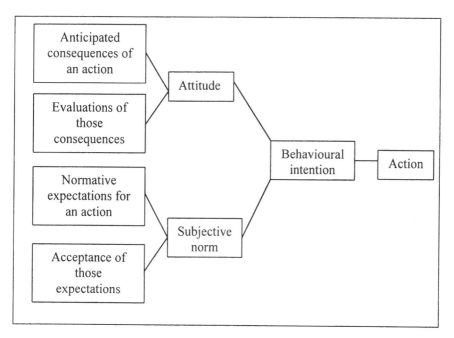

First, 'Attitude' is in most cases regarded as the dominant predictor of behavioural intention. This construct comprises the 'Anticipated consequences of action' and the 'Evaluation of those consequences'. In other words, 'Attitude' towards the action (i.e. favourable or non-favourable orientation) is predicted by the complex interaction of the effects of that action on the self and others and the perception of what tangible impact these actions will have. For example, voting Conservative might have the anticipated impact of a reduction in income tax. However, the individual must also evaluate the effect of this tax cut (for example, on the one hand more disposable income, on the other a possible cut in public services). The other predictor of 'Behavioural intention' is the 'Subjective norm' of behaving in a given manner. This term refers to the various social pressures under which the individual operates and is an expression of the degree to which significant others influence decision-making. This construct is comprised of two logical precursors: 'Awareness of norm to act' and 'Acceptance of norm to act'. In other words, the individual must be aware of social influences and pressures to do or not do a given behaviour *and* accept these pressures. A good example might be with kerbside recycling. A subjective norm in this context would operate such that if one was aware that those around were recycling and that one accepted this as 'normal' or 'normative' behaviour then the combination of these two factors would lead to modification of the intention to behave.

This theoretical framework for examining and predicting all types of social behaviour is still used today (e.g. Goldenhar and Connell, 1992-1993). However, one important addition has been made to the framework since its inception in the 1970s. Ajzen and Madden (1986) have amended the model to add one further predictor variable of behaviour. An acknowledgement by Ajzen and Madden that perception of the ability to act could be important in shaping intentions to act led them to add 'Perceived behavioural control' as a further predictor of behavioural intention as well as a direct predictor variable of behaviour, thus forming the *Theory of Planned Behaviour* (TPB). This self efficacy was examined in more detail in Chapter Three. However, suffice to state here that there are many occasions where individuals might be willing and intend to act, but do not. This new variable accounts for this occurrence.

Application of the TRA and TPB in Environmental Research

There has been very little use of the theoretical model (or any social psychologically grounded attitude-behaviour framework) outlined above in UK environmental studies (Canter and Donald, 1987). Whilst work exists concerning other aspects of the environment-behaviour relationship that have used such a model (Canter and Donald, 1987) the investigation of environmental behaviour and specifically waste management behaviour, is lacking. This may relate to rigidities in the disciplinary environment of the UK academic system (Kitchin, *et al*., 1997). Since the TRA is fundamentally a psychology-based concept, it has been left to psychologists to utilise. However, those interested in environmental behaviours within the UK, for example environmental scientists and geographers, have not used such techniques (see Chapter One). The following short review examines the efficacy of the TRA and TPB in explaining environmental behaviour.

Specific Environmental Studies Using the TRA and TPB

Many studies have sought to use the TRA and TPB within their broad research strategies and have used the framework either in its simplest form or as a direct test of the theory. Olsen's (1981) study of attitudes towards energy conservation is a good example of testing the TRA. Using data from three research projects, the various components of the TRA were assessed. Awareness of the positive consequences of saving energy and acceptance of the social desirability of doing so, was as important as the awareness and evaluation of the impact of saving energy. As Olsen (1981, p. 126) states:

> It might be argued that this expressed desire to contribute one's part to solving the energy problem is only a rationalisation for acting in one's own self interest, but that is irrelevant for the policy concern of how to promote energy conservation, since people do act in response to what they experience as internalized social norms.

In other words, it is difficult to distinguish the origin of the social norms, but nonetheless, it appears that they have an effect. Olsen raises a further point when discussing his results, asking the question:

> Do interpersonal pressures and situational contingencies affect one's conservation actions regardless of one's intentions? (Olsen, 1981, p. 127).

As will become increasingly apparent, the inability of the TRA and TPB to incorporate these variables adequately is a major problem of the framework.

Jones (1990) also used the TRA in his analysis of paper recycling in a faculty of a north-western university in the United States. Measuring the principal components of the TRA in Figure 5.1 above, Jones found strong links between the constructs, in particular the intention to behave and behaviour, as well as the combined effect of attitude and subjective norms and intention to behave. However, in this case, regression analysis demonstrated that unlike the Olsen (1981) model, attitude had a higher predictive value for intention to behave than subjective norms. Hence, behavioural intention derived more from the anticipated consequences of the action than from social pressure.

Work by Kok and Siero (1985) at Zeist, in The Netherlands, demonstrated how different factors can impact on the mechanics of the TRA in different ways. By asking respondents about their tin can recycling behaviour, they used the TRA as recommended by Fishbein and Ajzen, but with the addition of variables measuring the amount of difficulty encountered with the recycling. Although attitude and subjective norms did correlate with behavioural intention and behaviour itself, problems of the logistics of tin can cleaning were as important. Hence, only partial support for the model was given.

This partial support was confirmed by a study of attitudes towards water conservation by Kantola *et al.* (1982) who used the TRA to implement a questionnaire at a County Fair in the United States. They found that social influence was a good predictor of behavioural intention. However, age also had a significant bearing on the intention of respondents to conserve water. As Kantola *et al.* make clear, this is certainly at odds with the TRA. Goldenhar and Connell (1992-1993) examined first year student attitudes and behaviours towards paper recycling in a Mid-Western university in the United States. Again, statistical analysis demonstrated the power of the model to predict behaviour from the constructs in the TRA. However, the authors added two additional factors into the model to assess their effect. They found that past experience of recycling had almost as large a direct effect upon behaviour as behavioural intention and that this varied according to gender, with females' past experience providing higher predictive value for behaviour from past recycling experience. This impact of previous behaviour had also been found by Macey and Brown's (1983) analysis of energy saving behaviour, where experience was the highest predictor of behaviour! As Olsen (1981) had surmised previously, other factors are indeed important in predicting behaviour.

Ajzen and Madden's (1986) and Ajzen's (1991) amendment to the TRA to form the TPB was a step forward and in the two experiments undertaken to test the framework (outlined above) it was found that perceived behavioural control provided a more complete predictive framework. However, as mentioned above, unlike attitudes or subjective norms, this self efficacy or volitional control could act to predict both the intention to behave and behaviour itself, depending upon the situational circumstances involved. Chan (1998) lends support to the Ajzen model in his study of recycling in Hong Kong. As with Olsen (1981) and Jones (1990) above, attitude was more important a predictor of behavioural intention than subjective norms. Indeed, as predicted by Ajzen, perceived behavioural control was significant when predicting behavioural intention. Nevertheless, the high correlations found in previous studies between behavioural intention and behaviour itself were not found. Chan reported a Pearson correlation coefficient (r) of only 0.38, whereas Jones reported an r of 0.69. As Chan (1998, p. 325) points out:

The results indicated that Hong Kong people are only paying lip service to environmentally responsible actions. They intend to participate in environmentally protective behaviour, but do not pay effort and time in storing and sorting wastes.

Other analysis of the TPB has been even less supportive. Boldero's (1995) analysis of first year psychology students at an Australian university and their role in recycling behaviour indicated that when the TPB is placed under stringent statistical analysis, the principal components are less important. Using bivariate analysis, the principal components of the TPB were all important in some way in predicting those who did and did not recycle. However, multivariate analysis revealed that although the link between intention to act and behaviour was good, there was little effect from perceived behavioural control or the attitude and subjective norm factors. Not only were other factors related to intentions to behave, but also for those whom the attitude-behaviour link was weak, very little explanation was given. Boldero (1995, p. 458) concludes:

> ...other factors must be taken into account if we are to understand why individuals do not carry out their intentions.

Hence, as has been alluded to above with reference to other studies, the TRA and TPB both suffer from the fact that other variables are involved in the explanatory process and that the attitude-behaviour link is not certain.

Taylor and Todd (1997) partly reinforce this conclusion with their results from a study of residents in a medium sized city in Canada. They tested three models of waste management. First, the TRA was examined. Second, a 'Belief-behaviour' model and third an 'Integrated Waste Management' model. They found that although the TRA gave a good result for the intention-behaviour consistency, the prediction of intention to act was poor, with subjective norms insignificant. Their integrated waste management model gave improved results as a consequence of aggregating the stronger points from the TRA and 'Belief-Behaviour' model. Continued research by other workers suggests that other case-specific variables can be important in predicting behavioural intention when added to the framework. For example, in Lam's (1999) study of water saving, the addition of 'perceived moral obligation to act' and 'perceived water right' predicted behavioural intention. Similarly, Kaiser *et al.* (1999) analysed environmental knowledge and values and found these to be excellent predictors of their general ecological behaviour measure.

Efficacy of the TRA and TPB

This brief review of the literature examining the TRA/TPB is limited to those studies within the realm of environmental behaviour. However, the problems that have been identified have not been confined to the environmental literature. Both Liska's (1984) and Bagozzi's (1992) reviews of attitude theory question the efficiency of the TRA/TPB in terms of the extent to which 'other factors' can be constantly added and modified. Hence, although the Fishbein-Ajzen framework may be a useful framework for research, it is obvious from the review given here that the model itself has limitations when examining specific behaviours. Indeed, it seems sensible that for specific behaviours, 'other factors' would be involved in shaping behavioural intention and behaviour. Additionally, although it might be the case that 'behavioural intention' is important and reflects the relationship between oral acceptance of behaviour and actual action, it cannot be assumed that the only predictor of behaviour is verbal intention. Thus, although the TRA/TPB provides a logical model, a framework of environmental behaviour would a more complex framework that can both incorporate as many of the important variables that are likely to affect behavioural intentions and behaviour, as well as assessing the relative impacts of these variables upon the intention-behaviour relationship. Hence, as has been demonstrated by this short review, whilst a 'model' of behaviour has utility, the irrationality of individual action means that the 'framework' approach, whereby factors can be added or omitted and the structure of a process altered, has far greater utility in examining variance in behaviour.

A Framework of Environmental Behaviour

The impelling logic of the TRA nonetheless provides the basis for a framework of environmental behaviour. The link between intention and behaviour provides a useful basis on which to examine the disparity between stated intentions and behaviour. Nonetheless, the need to examine the efficacy of the three sets of variables previously described is of importance within this context. Figures 5.2 shows a framework of environmental behaviour.

As can be seen, behavioural intention and behaviour form the central portion of the framework, with the three sets of factors placed on the periphery. Environmental values are placed to the left of the centre because the research outlined in Chapter Two indicated that there is equivocal evidence concerning the impact of such values on behaviour. A number of alternative reasons for this variation have been offered, including both theoretical and methodological factors. It is argued here that although the exact reason for this variation is as yet unclear, environmental values might be indirect precursors to behaviour. Nonetheless, it is argued that such values are the logical antecedents to pro-environmental action since a more ecocentric value base would lead to positive intentions towards action.

Situational and psychological factors are placed at mutually peripheral points of the framework. It is argued that alternative factors from each set may indeed influence both intentions and behaviour.

Such a conceptualisation and framework of behaviour provides the flexible and loose structure that permits the addition or omission of variables as appropriate. Yet there is still some form of structure that can be 'tested'. Nonetheless, by adding flexibility, this testing is not a matter that could enable the framework to 'fail'. As such there is no process implied in the sense that the only logical requirement is that there be a satisfactory causal relationship between intention and behaviour. As can be seen from the detail in the framework, the individual variables are added as and how the individual author prefers, but are given equal dominance in causal terms. Although based around the TRA, the framework is a significant step forward for social scientists interested in examining environmental behaviour. One of the most compelling and difficult questions to answer in recent decades has been the reasons for stated aspirations to protect the environment not being reflected in individual actions. This framework provides a means by which to conceptualise this relationship whilst taking into account the influencing factors of intention and/or behaviour. As such, it handles both the 'value-action gap' problem, as well as providing a way of examining numerous variables within the context of each other.

Applying the Framework: Waste Management Behaviour

The framework therefore provides a unique conceptualisation of environmental behaviour based on a logical model, a review of theoretical and empirical literature and previous frameworks. The structure can be used to examine any environmental behaviour or more than one at a time. Although the framework in Figure 5.2 is merely a conceptualisation, there is great scope for applying such a framework to an individual study in terms of planning, implementation, analysis and interpretation. This is described in Chapters Six to Eight where the framework is used as the basis for a study of individual attitudes and behaviours towards waste management options in Exeter, UK. The reader is shown how the study is initiated, planned and implemented. Analysis tools and techniques are described and the data are interpreted using the conceptual framework given in Figure 5.2. Finally, the policy implications and prospects for using such a framework in studies of other environmental behaviours are examined.

Figure 5.2   Conceptual framework of environmental behaviour

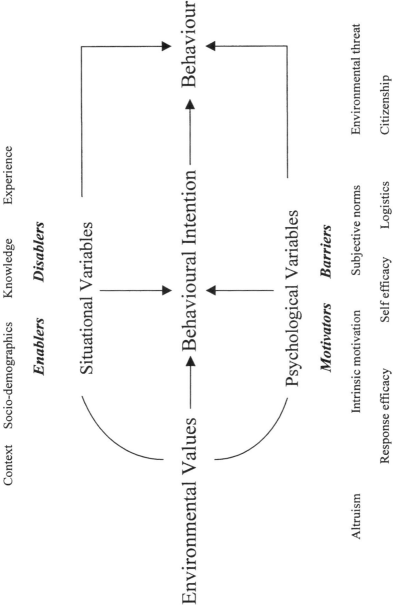

6 The Exeter Study I: planning, implementation and descriptive statistics

Introduction

This chapter demonstrates how the conceptual framework developed in the preceding chapter can be used to plan and implement a study of attitudes, behaviours and linking variables. As stated before, the specific example used here is household waste management, but there is no constraint on adapting the methodology described here in order to study alternative environmental behaviours. The study is initially set in context and the research area of Exeter is introduced. The development of a survey instrument is then outlined and explained. The implementation strategy is then examined. Finally, the descriptive statistics of the study are presented and the validity of the study evaluated.

The Exeter Study in Context

The Exeter study was undertaken in order to test the overall efficacy of the conceptualisation presented in the previous chapter. As such, the design of the research was focused around this framework. The specific study had three objectives:

- To describe the attitudes and behaviours of a representative sample population with regard to household waste management;

- To explain the patterns of behaviour as far as possible within statistical limits; and

- To make both theoretical and policy-related recommendations on how environmental behaviour could be better understood and what actions local authorities could take to enhance waste management behaviour.

As such, the study required a suitable sample population for which the characteristics specified with the conceptual framework were covered and data with which to conduct suitable data analyses and, more importantly, to make reliable recommendations with.

Bearing these requirements in mind, a strategy was developed that focused on these salient necessities. It should be borne in mind, however, that the study presented here is an example of how such a such a study can be undertaken based on a conceptual framework of the type given in Chapter Five. Local circumstances will of course dictate the actual study design, but the description of this design that follows is a practical worked example of how such a study can be undertaken.

Study Area: the City of Exeter

Exeter is a Cathedral city located in the South West of England. The city has a population of approximately 90,000 which is focused mainly on the eastern side of the River Exe. This upland side of the city comprises the centre and major housing developments, whilst there is significant industrial development on the floodplain to the west of the river, along with several other housing districts. Housing is primarily terraced or semi-detached, with few high rise apartment blocks on the one hand, or few executive homes on the other.

The city currently has a good recycling rate of 21.8% (Exeter City Council, 2000a) and is in the process of opening a new materials reclamation facility (MRF) near the city to handle increasing volumes of recyclable waste. The City Council, as the Waste Collection Authority (WCA), operates a 'Recycle from Home' scheme which currently covers half of the city's households. In this scheme, residents have two bins; a green bin, collected fortnightly, which takes recyclable material, including paper, magazines, steel cans, aluminium cans, plastic bottles, but not glass; and a grey bin, also collected fortnightly, which takes all other refuse, to be landfilled. The other 50% of residents can recycle waste at a number of static recycling 'bring' sites located at civic sites, car parks, shopping centres, etc. These static recycling sites offer a range of recycling opportunities, although spatially this is variable. Still by far the easiest means of recycling most recyclable material is in the 'Recycle from Home' bin.

The city therefore presented an ideal opportunity to examine the spatial and contextual elements of waste management as emphasised in the conceptual framework. However, use of one urban area for such a study, whilst very common in this type of research, is not necessarily ideal. This study was undertaken in one urban area due to significant fiscal and logistical constraints. Yet, a study that could incorporate a rural dimension and a wider socio-demographic base might be more suitable. On the other hand, such a study loses some of the valuable insights offered by examining just one specific area and the important policy outcomes that can be achieved. Whichever is chosen, there must be enough variation in the variables within the conceptualisation used to allow for statistical testing. Thus, many levels of kerbside provision would be difficult to accommodate without a large sample, since the divisions would make statistical significance harder to achieve in multivariate analyses.

Study Instrument

Having chosen a specific study area, there is the question of the type of study instrument that should be used. The case has been compellingly made throughout this book that standardised, quantitative and statistically rigorous data are of greatest value to such a study and therefore there is little option but to use a questionnaire of some description. This could be of any type, from a telephone survey to a mail questionnaire. The choice is based on a number of factors, such as time, money available, labour and the need for a good response rate. In the Exeter study, the need for a high response rate was of paramount importance and so, taking into account the favourable labour situation and limited financial resources, a door-to-door survey was devised, where the questionnaire could be left for the householder to complete and, after a few days, be collected by the researcher.

Respondent completed questionnaires have to fulfil a number of key requirements, not least that they are aesthetically appealing and easy to complete. The Exeter questionnaire was designed using the recommendations of Dillman (1978) in his 'Total Design Method' where he emphasises the need for presentation, ease of questioning and a significant play on the importance of motivational sentences that encourage participation. This should, at all costs, avoid phrases such as 'this is for my PhD thesis'! The destination of the results after they have been analysed is also important to emphasise.

Bearing these considerations in mind, the questionnaire used in the Exeter study had a bold front cover, featuring the Agenda 21 logo for Devon, with the phrase 'A Better Devon, A Better World' encircling the two globes. The survey was called 'Making Exeter Cleaner', a play on this desirable outcome which it was hoped would enhance response to the survey. On the first page, respondents were introduced formally to the questionnaire and informed of its importance and, most, crucially who was conducting the survey and how they could be contacted.

The subsequent ten pages were structured questions relating to the conceptual framework. The beginning of the survey attempted to be as unobtrusive as possible by requesting information regarding lifestyles. This forms the behaviour element in the conceptual framework and measured behaviour over a range of activities: waste minimisation (5 items), reuse (5 items) and recycling (10 items). These behaviours were selected from a list of 'Environmental Actions' posted on the City Council website (www.exeter.gov.uk) which the local authority seeks to encourage. The emphasis on recycling was for two reasons. First, this is an activity that is most widely associated with being pro-environmental or 'green'. Second, in order to have an objective measure of context, it was necessary to gauge responses from participants both in the 'Recycle from Home' scheme and from those residents who did not have this facility. The respondent scored each item according to the frequency with which that action was undertaken, from (1) Never to (5) always. All of the statements were personalised and the use of emphasis in sentences (e.g. 'very', 'always') was avoided to reduce the amount of possible confusion between statements.

The second section of the survey examined personal factors associated with socio-demographics within the conceptual framework. Age was categorised to increase response rate, with five possible responses. Gender was a tick M/F question. Car access was a tick Yes/No question. Household composition comprised a range of household types. Education was a hurdle question, whereby if respondents had completed compulsory education they were asked to tick the level they had achieved after this qualification. Occupational status was assessed by means of a series of broad categories that outlined the current status of individuals (employed, unemployed, unfit for work, retired, student). These helped assess the nature of the income data, which were grouped into four very broad categories, again to increase the number of respondents who answered this question. Finally, the controversial issue of political persuasion was included, where respondents were asked to state which political party they would vote for if there were a general election 'tomorrow'. Respondents were given both a 'Wouldn't vote' and 'Pass question' option. In addition, the researcher would measure variables such as house type using his discretion on delivery.

The third section of the survey was concerned with part of the 'knowledge' section of the framework, focusing on generalised knowledge about the environment and waste (or 'abstract' knowledge). This particular section assessed general environmental, waste and policy knowledge. Respondents were asked to score each statement according to whether they believed it to be true or false. The first ten statements were drawn from *Social Trends* (CSO, 1995; 1996; 1997; 1998). Dealing with 'environmental' and 'waste' knowledge, these questions were set on a true or false basis and used the statements of fact from *Social Trends* to examine individual awareness of global environmental issues in the first instance (such as global warming and ozone depletion) and waste-specific issues in the second instance (such as the problem of landfilling waste and the amount of waste recycled in the UK). The last two generalised knowledge questions were policy measures, where respondents ticked whether they were aware of 'Sustainable Development' and 'Local Agenda 21'.

The fourth section assessed the 'environmental values' part of the conceptual framework. For the purposes of the Exeter study, a new set of statements were constructed due to the inadequacy (Barr, 1998) of the most widely used alternative (the NEP). The statements gauged four constructs: ecocentrism, strong sustainability, weak sustainability and apathy. The first of these is the classical view of ecocentrism as presented by O'Riordan (1985), but the second, 'strong sustainability' represents a weaker view, focusing on environmental protection, but within a human setting. 'Weak sustainability' presents a more classically technocentric position, incorporating the notion of the insignificance of 'critical natural capital' (Pearce *et al.*, 1989). Finally, 'apathy' incorporates the view held by Thomson and Barton (1994) that indifference towards environmental issues provides an important construct.

The 'behavioural intention' scale (fifth section), the focus of the centre of the conceptual framework, related directly to the behavioural questions in the first section, but was disguised by the prefix: 'how willing are you to help the environment'. It was hoped that this would measure the desired variable (behavioural intention), but not make the link to the first section too obvious!

The sixth section concerned 'psychological' variables and inasmuch detailed statements that assessed subjective norms, responsibility, satisfactions, citizenship, logistics and so on. All were scored on a 5-point agreement scale.

Leading on from this, the seventh section sought information regarding sources of information on waste management (such as TV, newspapers, magazines and radio). This was a tick box question, where respondents indicated how many sources they used. The eighth section examined local waste (behaviour-specific or 'abstract') knowledge. Local waste knowledge was assessed by a series of yes/no questions concerning items that could be recycled in Exeter. In addition, those who had a 'Recycle from Home' bin were asked to tick items that could be recycled therein. The question was posed such that respondents were asked 'what' services they knew were available, rather than 'if' they knew that they were available. This was designed to prevent people merely ticking yes to impress the researcher and inasmuch a few rogue variables were added!

The penultimate (ninth) section dealt with experience and was assessed by means of asking respondents who did take part in the 'Recycle from Home' scheme whether they had undertaken any recycling behaviour before the introduction of the scheme. Respondents were finally asked whether they had any further comments and were thanked for their participation in the survey.

Study Implementation

Having designed the questionnaire instrument, a suitable means of implementing the survey with maximum efficiency and response rate had to be devised. As alluded to above, this was undertaken in this particular study by means of a door-to-door system, or as has been termed in previous accounts, the 'contact and collect' methodology. In this system, the researcher calls at a pre-determined address (see below) and introduces the survey. The respondent then fills out the questionnaire in his or her own time and it is collected by the researcher again in a few days. This has a number of advantages, not least the personal contact with the respondent and the explanation of any constraining issues. It also enables the researcher to collect further valuable information.

The selection of the sample that is used for the questionnaire is important. There is a need to select a sample that represents both the characteristics within the conceptual framework being used, as well as being representative of the population as a whole. In this instance, a probabilistic sampling technique, using randomised techniques was employed. Under this method, households are selected at random from the electoral register, a technique outlined by Gray (1971). By selecting the ideal sample size, the technique will provide almost the exact number of addresses required, but at random.

Of course, this textbook description of how to sample the population in question is academic in the real world. The pilot survey undertaken during the planning of this survey identified that contact was severely restricted when sticking rigidly to the households identified. Thus, a system was devised where the subsequent household was sampled until a response was obtained. The normal procedure would then be resumed.

Finally, as mentioned in the previous paragraph, a pilot study is essential in order that the specific sampling, questionnaire and other issues can be ironed out. In the Exeter study, the pilot questionnaire focused on a particular ward of the city and included a further short questionnaire asking people for their thoughts on the survey. This provided valuable information regarding the improvement of the study instrument.

A Note on Data Analysis

Thus far in this book there have been numerous references to quantitative data analysis techniques and statistical justification for patterns and trends. It is now prudent to examine these techniques and strategies in more detail before exrmining the dynamics of the Exeter study in detail.

The first crucial point to note here is that whenever a questionnaire of this type is being planned, it is vital to examine the extent to which the information sought can actually be found from the data collected. Indeed, from a personal point of view, it is also necessary to gauge the extent to which the researcher is capable and comfortable with, for example multiple non-linear regression using log-it analysis – certainly not taught in most geography departments!

The structure of data analysis will be of vital importance in order to move from basic description of data, partly to examine issues of representativeness and become familiar with the data, and partly to examine the data types, for example normal and non-normal distributions and issues of skewness and kurtosis. From the point of data description, basic inferential statistics are needed to examine the possible emergent patterns within the conceptual framework being used. Such 'bivariate' statistics include Pearson's correlation coefficient (r) (for parametric data), Spearman's rank correlation coefficient (r_s) and the Chi-Square (χ^2) test (for non-parametric and nominal/ordinal data sets). These tests are explained in more detail when they are utilised in Chapter Seven below. Such tests are based on the scientific method of hypothesis testing (Moore and Cobby, 1998) whereby null and alternative hypotheses are formulated and examined on the basis of theoretical experience. Thus, although not made explicit in the following chapter, each statistical test is grounded in testing the null hypothesis (H_0) that, for example, there is no relationship between behavioural intention and behaviour. Acceptance of the alternative hypothesis can only be made where a statistical test, such as a Spearman Rank test, shows that there is a significant relationship between intention and behaviour.

Having examined the bivariate relationships between the variables in the conceptual framework, analysis must proceed to data explanation. This has a number of steps. Initially, the empirical dimensions of the data collected need to be examined. It is likely that the behavioural data, behavioural intention, values and psychological data, having been measured on ordinal scales, can be aggregated and examined in relation to their empirical correlation such that a set of interval-measured factors can be formed as scales to be used in multiple regression analyses, a fundamental assumption of which is that scales are normally distributed. Factor analysis permits both data aggregation, whilst at the same time an examination of the degree to which items 'load' with each other to make more logical sense. For example, a 'recycling behaviour' factor may emerge, or an 'ecocentric' value factor may become apparent.

Having created these factors, they can be used in multiple regression analyses in order to examine the determinants of the dependent variables within the conceptual framework. Multiple regression is a procedure whereby the linear relationship between independent variables and the dependent variable can be predicted by use of a 'regression equation', an expression of the value of the dependent variable when the independent variables are at a certain level. Techniques such as stepwise regression now permit the best possible sub-set of regressors to be found with ease on computer packages such as MINITAB. Indeed, although not dealt with here, it is possible to develop a fuller understanding of the determinants of independent variables by means of non-linear regression models (Lea, 1998).

The flow diagram below (Figure 6.1) demonstrates a classical data analysis programme that can be used with such data. The net result of the data analysis programme is the construction of models of each behaviour, by representing the regression results by means of path diagrams which can be used to show the study findings in the same format as the conceptual framework, which aids interpretation of the data for both academics and policy makers.

Descriptive Statistics

Sample Response and Representativeness

The Exeter study was undertaken between September and December 1999 and of 983 address selected, full responses were gained from 673 households, representing a 69% response rate. This was an excellent result and enabled the results produced from the data to be given extra support.

In terms of the representativeness of the wider Exeter population in demographic terms, the sample was on the whole acceptable. Using the technique adopted by Oskamp *et al.* (1995), the demographic breakdowns were compared to the latest available Census data. In all cases the data for age, gender, car access, household type, occupation and political affiliation were within 10% of the Census data (OPCS, 1991-92). In the cases of house type, income and education, no Census data was available.

Figure 6.1   Data analysis programme

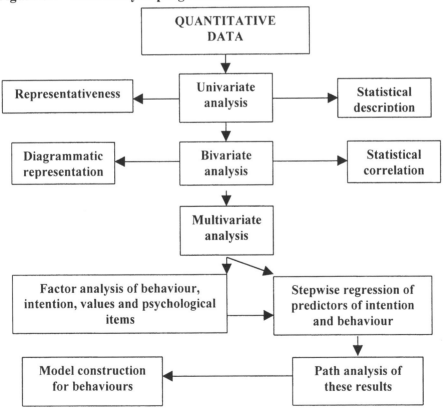

Waste Management Behaviour and Behavioural Intention

Figures 6.2 and 6.3 below show the relative scores for the behaviour and behavioural intention items as in the questionnaire survey. The first point to note from Figure 6.2 (behaviour) is that there is a significant difference between recycling and the other behaviours. Recycling is very definitive behaviour that is characterised by keen participation or no involvement. Few respondents reported that they 'sometimes' recycled. Given the large amount of respondents who stated that they 'always' or 'usually' recycled, recycling is clearly a well-established activity amongst many of the residents in Exeter. Yet there is a significant number who never undertook such activities. This is in contrast to reuse and, to a certain extent, minimisation behaviour. Whilst far fewer people undertook these activities as a matter of course, most people at least 'sometimes' took part. There does, on the whole, appear to be a significant difference between the recycling and other behaviours and between minimisation and reuse.

Figure 6.3 shows the scores for the same behavioural items, but with regard to a willingness to undertake each action, the variable used in this study for 'behavioural intention'. What the figure shows is fascinating. First, the difference between recycling behaviour and the other actions is still apparent. People were generally very willing to recycle items. Nonetheless, there were still a significant number who stated that they were very unwilling. This is in contrast to the situation with minimisation and reuse behaviour, where virtually no one stated that they were very unwilling to act. In this case, most were willing to minimise and reuse. Second, and related to this point, is the fact that in all cases peoples attitudes were generally more favourable to acting than their behaviour suggests. This justifies the behavioural intention question as it is important to note the difference between stated behaviour and willingness to act.

Overall, the behavioural and intention data suggest that there is a disparity between intention and action, but that this is linked. Whilst the individuals within the sample had definitive recycling attitudes and behaviour, they were more equivocal both behaviourally and intentionally concerning minimisation and reuse.

Environmental Values

Figure 6.4 shows the scores for each environmental value item. As shown in the legend of the chart, four of the items were recoded to give equal weighting to all statements, enabling all higher scores to equate to more positive environmental attitudes.

Figure 6.2 Waste management behaviour (Note: reported behaviour. See text on Page 96)

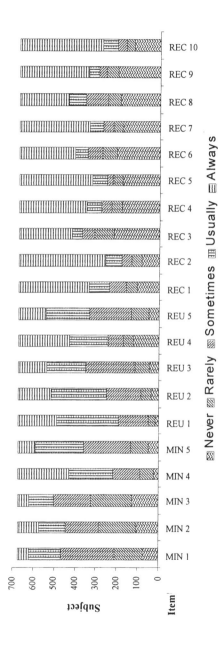

■ Never ▨ Rarely ▨ Sometimes ▥ Usually ▤ Always

Item explanation:

MIN 1   Buy produce with as little packaging as possible (2.97)
MIN 2   Use my own bag when going shopping, rather than one provided by the shop (2.91)
MIN 3   Look for packaging that can be easily re-used or recycled (2.67)
MIN 4   Buy fruit and vegetables loose, not packaged (3.88)
MIN 5   Buy products that can be used again, rather then disposable items (3.38)
REU 1   Try to repair things before buying new items (3.94)
REU 2   Reuse paper (3.72)
REU 3   Reuse glass bottles and jars (3.44)
REU 4   Wash and reuse dishcloths rather than buying them new (3.5)
REU 5   Reuse old plastic containers, like margarine tubs (3.48)

REC 1   Recycle glass (3.78)
REC 2   Recycle newspaper (4.05)
REC 3   Recycle food cans (3.06)
REC 4   Recycle drinks cans (3.54)
REC 5   Recycle junk mail (3.56)
REC 6   Recycle foil (3.19)
REC 7   Recycle cardboard (3.52)
REC 8   Recycle textiles (3.21)
REC 9   Recycle plastic bottle (3.39)
REC 10 Recycle magazines (3.86)

Scored on a five-point Likert scale (1 = never to 5 = always). Means are given in brackets alongside each statement

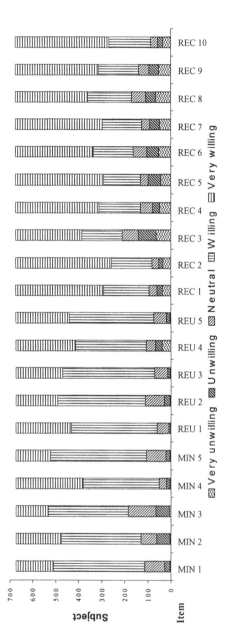

Figure 6.3   Behavioural intention

Item explanation:

MIN 1   Reduce the amount of produce that's bought which has lots of packaging (4.05)
MIN 2   Take old plastic bags shopping, rather then using new ones, or take a durable bag (3.98)
MIN 3   Look for wrapping that can be easily reused or recycled (3.82)
MIN 4   Buy certain produce without packaging, like fruit and vegetables (4.32)
MIN 5   Buying fewer disposable products (4.03)
REU 1   Repairing items before deciding they have to be thrown away (4.25)
REU 2   Reuse paper, rather then buying it new (4.05)
REU 3   Reusing jars and bottles wherever possible (4.17)
REU 4   Wash and reuse certain items before disposing of them, like dishcloths (4.07)
REU 5   Reuse old containers, like ice cream tubs or margarine boxes (4.19)
REC 1   Recycle glass (4.28)
REC 2   Recycle newspaper (4.36)
REC 3   Recycle food cans (3.81)
REC 4   Recycle drinks cans (4.13)
REC 5   Recycle junk mail (4.15)
REC 6   Recycle foil (4.02)
REC 7   Recycle cardboard (4.16)
REC 8   Recycle textiles (3.95)
REC 9   Recycle plastic bottles (4.1)
REC 10 Recycle magazines (4.37)

Scored on a five-point Likert scale (1 = very unwilling to 5 = very willing). Means are given in brackets alongside each statement

There was strong agreement with the notion of 'strong sustainability', indicating that most respondents believed that there are critical natural limits that must be adhered to if both the natural and human environment is to be preserved for future generations. Of interest also is the fact that there was strong agreement with the first 'ecocentric' item, dealing with the value of nature and humans. Over three-quarters of respondents evidently believed that humans and nature have equal value. This is a fascinating finding, since it has been argued that such notions are the reserve of a much more select segment of the population (O'Riordan, 1985).

Very strong agreement also existed with regard to the notion that human changes do have significant environmental impacts (Statement WS1). This implies that there is at least an awareness of some human effect on the environment. Indeed, this would appear to be represented in the amount of concern shown (Statements AP1 and AP2).

Figure 6.4   Environmental values

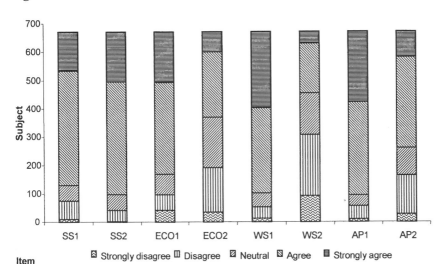

Item	
SS 1:	The environment is forgotten too often when decisions are made (3.88)
SS 2:	If we over-use our natural resources, human development may be harmed in future (4.04)
ECO 1:	Nature and the environment have as much value as human beings (3.8)
ECO 2:	Humans should not develop any more resources or land, in order to protect the natural environment (3.21)
WS 1:	Nature isn't harmed by human changes* (4.14)
WS 2:	Human welfare should be our primary concern in the future* (2.92)
AP 1:	The environment is of little concern to me* (4.12)
AP 2:	Getting through daily life and surviving is what concerns me the most, not the environment* (3.45)

* Reverse coded
Means are given in brackets alongside each statement

However, these positive findings are overshadowed somewhat by the results for the second 'ecocentrism' statement. Evidently, well under half the respondents felt that human development should cease to preserve the natural environment. This is interesting, since respondents had so enthusiastically endorsed the notion of nature and humans being equal. Indeed, the second 'weak sustainability' statement (recoded) implies that two thirds believe that human welfare should be the primary concern in the future.

Of course, the author defined these statements, and as will be seen in Chapter Seven below, the empirical dimensions of these statements are not as initially anticipated. However, it is of great interest that although respondents acknowledged human exploitation of the environment, appeared concerned, and even felt nature had equal value with humans, they did not feel that this should halt development or reduce the priority given to human welfare. In a sense, there is the feeling that some items were emotionally simpler to complete than others. This is an important finding, since it implies that there may be certain levels of acceptable environmental concern, for example stated values of nature and humans, as opposed to human development.

Situational Variables

Kerbside Recycling Facilities 54% of the sample had access to the 'Recycle from Home' scheme. This was representative of the population as a whole where 50% of the city is covered by the scheme. No comparable data are available for static services, but as noted above, each respondent received a 'recycling score' on the basis of the number of materials recyclable at their nearest bring site. These ranged from one to seven items.

Socio-demographics See the section above on sample representativeness.

Knowledge In terms of general environmental, waste and policy knowledge, Figure 6.5 shows the percentage of respondents answering each item correctly. Knowledge of environmental issues did not fall below 70% and in two cases (ozone depletion and road traffic emissions) the percentages were even higher. In relation to knowledge about the waste problem, the scores were reasonably high, but most interesting is the clear knowledge people have of the lack of space in which to landfill rubbish.

However, the impact of this result may be somewhat muted, since only just over 50% of respondents answered the previous statement correctly, namely that they thought that landfill was a small part of the way waste was disposed of. Hence, although people may know landfill is at a premium, this would not matter in terms of the waste problem since they do not believe it is crucial to waste management. Knowledge of the waste produced was, however, reasonably good, with around 70% answering correctly to questions regarding the origin of municipal waste. The fact that a similar number thought that recycling was a small part of the waste disposal process indicates that there may be ambiguity amongst respondents concerning knowledge about where and how their waste is dealt with, since many did not believe landfill was important either. The perception of the disposal process is an issue that has not been dealt with in the structured survey, and will be followed up in future chapters. However, the fact that few respondents believed landfill was important is fascinating, and will be examined in the next chapter with regard to other variables.

Figure 6.5   Percentage of correct answers to knowledge questions (N 670)

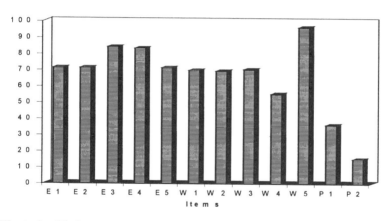

E1: Nine tenths of the hottest years on record have been since 1983
E2: Warming in the atmosphere is due mainly to increases in human produced carbon dioxide
E3: CFCs in aerosols and other packaging cause the depletion of the ozone layer
E4: Road transport produces few harmful emissions in to the atmosphere*
E5: One-third of total household water supply is used to flush the toilet
W1: Households produce a minority of the waste handled by councils*
W2: Each average household produces 21 kilogrammes of waste every week
W3: Under one-tenth of household waste is recycled
W4: Sending waste to a landfill site is only a small part of the way
      waste is disposed of*
W5: Landfill sites are running out of space for more waste
P1: Have you heard of 'Sustainable Development'
P2: Have you heard of 'Local Agenda 21'
* Indicates that a respondent would receive a point if they answered 'no' to this item

Nevertheless, the fact that the environmental knowledge scores are similar to the waste knowledge scores means that waste is part of people's knowledge base just as the more 'popular' global issues are. This does not hold, however, when the policy items are examined. Respondents were asked to indicate whether they had heard of two key policy frameworks that are seen as key to achieving sustainable waste management in the UK. Although a third of respondents were aware of sustainable development (and that does not imply they know what it is) only a quarter had heard of Local Agenda 21, which seems surprising, since the city produced its LA21 over two years ago, supposedly 'written' by the people of Exeter. Such a finding supports, in crude form at least, the notion that LA21 is still the preserve of what Selman (1996) has termed the 'champions' of local environmental planning. The impact of this lack of knowledge will be examined in the forthcoming chapter, since it is hypothesised that those more actively involved in such campaigns, and those who at least have heard of them, are more likely to participate in pro-environmental actions, such as waste minimisation, reuse and recycling behaviour.

Local (behaviour specific or 'concrete') waste knowledge is defined here as that knowledge that respondents have of the recycling facilities that are available in the city. For those with the 'Recycle from Home' scheme, this also refers to the materials that can be placed in the recycling receptacle for fortnightly collection. The percentages for each item are given in Figure 6.6 below.

Remembering that certain items cannot actually be recycled in Exeter, and were added as check statements, it can be seen that knowledge was far better among those residents with a recycling bin at home. As might be expected, there are higher scores for what has been termed the 'classic' recyclables (glass, cans, and newspaper), and lower scores for 'marginal' items that are not as popularly advertised, such as textiles, cardboard, and plastic bottles. Of course, it should be expected that this knowledge relationship is not spatially uniform over the city, and as such the following chapter will examine the relationship between spatial provision and knowledge of recyclable materials.

Of the check items that cannot be recycled in certain ways, it can be seen that one tenth of respondents thought glass could go into their 'Recycle from Home' bin, a problem which the local authority has highlighted. Due to the manual sorting of recyclables, the Council does not permit glass in the 'Recycle from Home' scheme. However, more worrying from a contamination point of view is that 75% of residents with a 'Recycle from Home' bin thought that other plastics could be deposited in the recycling bin. This is a major problem for the Council, since only plastic bottles are permitted. However, the consistent scores of over 90% for most of the other items for the 'Recycle from Home' scheme indicate that residents were familiar with the system. One exception is textiles. This may not be a problem, since such items are not recycled as frequently as other household goods, and the wealth of charity shops and clothing collections may explain this gap in knowledge, since people do not need to know that textiles can go in their green bin.

Figure 6.6   **Knowledge of recyclable items in Exeter. Per cent answering yes. ('Exeter' scores denote knowledge of city-wide facilities, N = 673; 'Home' scores are knowledge of 'Recycle from Home' facilities, N = 365).**

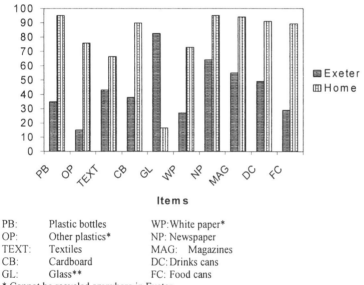

PB:	Plastic bottles	WP: White paper*
OP:	Other plastics*	NP: Newspaper
TEXT:	Textiles	MAG: Magazines
CB:	Cardboard	DC: Drinks cans
GL:	Glass**	FC: Food cans

* Cannot be recycled anywhere in Exeter
** Cannot go in 'Recycle from Home' bin

In summary, knowledge of respondents appears to be consistently high in environmental and waste management terms. Indeed, specific knowledge was good of the 'classic' recyclables. However, there are gaps in three areas. First, the disposal of waste may be ambiguous to respondents. Not knowing the problems faced as landfill is exhausted may be a significant gap in knowledge that could impact upon attitudes and behaviour. Second, respondents showed a lack of awareness of two key policy tools ('sustainable development' and 'Local Agenda 21'). This may impact upon relating the pro-environmental message of these campaigns, and practical shifts in attitudes and behaviour towards waste. Third, there is still a deficit in knowledge concerning the 'marginal' recyclables, and as such a need to focus more strongly on these items. In all three cases, the analysis must examine the extent to which knowledge of this kind impacts upon attitudes and behaviour.

Experience Of those with a kerbside collection of recyclables, 60% stated that they had recycled before the introduction of the kerbside service. This enables good comparison of the behaviours and attitudes of those who had and did not have experience of previous recycling and their possible response to this.

Psychological Factors

The large number of statements used to measure the various factors described in the lower section of the conceptual framework cannot be examined in great detail due to lack of space. However, Table 6.1 shows the concepts being measured, the statement type and the mean score for each item. As described above, two statements for each concept were used in order to provide a check system. The grouping of some statements is different on occasion to those within Figure 5.2 due to some overlap and ease of presentation, especially where the complex Schwartz model of normative influences on altruism is concerned.

Although means evidently do not provide the same detail as graphic distributions, it can be seen that scores on the items were generally positive and indicated overall support for a number of the concepts. In terms of analysis, there is little that can be said until such data are placed in the context of the behavioural and intention data. Suffice to state that responses were definitive and there were few 'neutral' responses.

Conclusion

This chapter has demonstrated how a study of attitudes and behaviours can be planned, implemented and used to gain data concerning many aspects of waste management. Yet this data are somewhat limited unless more complex analyses are undertaken. Having now examined the nature of the data collected, it is necessary to move on to examine the reasons for the behavioural patterns observed and how such patterns can be re-conceptualised.

Table 6.1 Psychological variables in the questionnaire

Statement type and concept	Concept(s) in conceptual framework	Mean
Perceptions of the waste problem		
WASTE PROBLEM PERCEPTION 1	Altruism - 'Awareness of need'	4.26
WASTE PROBLEM PERCEPTION 2	Altruism - 'Awareness of need'	4.04
WASTE CONCERN 1	Altruism - 'Personal norms'	3.8
WASTE CONCERN 2	Altruism - 'Personal norms'	3.63
THREAT1	Altruism - 'Awareness of need' Environmental threat	3.66
THREAT2	Altruism - 'Awareness of need' Environmental threat	3.53
Ascription of responsibility for waste		
ASCRIPTION OF RESPONSIBILITY 1*	Altruism - 'Ascription of responsibility' Citizenship	4.17
ASCRIPTION OF RESPONSIBILITY 2*	Altruism - 'Ascription of responsibility' Citizenship	3.94
OBLIGATION TO ACT 1	Altruism - 'Personal norms'	4.03
OBLIGATION TO ACT 2	Altruism - 'Personal norms'	3.69
Beliefs about waste management		
BELIEFS ABOUT WASTE MANAGEMENT 1	Altruism - 'Relevant actions available'	4.25
BELIEFS ABOUT WASTE MANAGEMENT 2	Altruism - 'Relevant actions available'	4.26
BELIEFS ABOUT WASTE MANAGEMENT 3	Altruism - 'Relevant actions available'	4.04
BELIEFS ABOUT WASTE MANAGEMENT 4	Altruism - 'Relevant actions available'	3.94
RESPONSE EFFICACY 1	Response efficacy	3.86
RESPONSE EFFICACY 2	Response efficacy	3.46
Logistics		
SELF EFFICACY 1	Self efficacy	3.12
SELF EFFICACY 2	Self efficacy	3.28
TIME 1	Altruism - 'Assessment of costs' Logistics	3.71
TIME 2	Altruism - 'Assessment of costs' Logistics	3.52
RECYCLING CONVENIENCE 1	Altruism - 'Assessment of costs' Logistics	3.29
RECYCLING CONVENIENCE 2	Altruism - 'Assessment of costs' Logistics	3.12
STORAGE SPACE 1	Altruism - 'Assessment of costs' Logistics	2.82
STORAGE SPACE 2	Altruism - 'Assessment of costs' Logistics	2.92
Subjective norms		
AWARENESS NORM TO RECYCLE 1	Subjective norms	3.27
AWARENESS NORM TO RECYCLE 2	Subjective norms	3.12
ACCEPTANCE NORM TO RECYCLE 1	Subjective norms	3.55
ACCEPTANCE NORM TO RECYCLE 2	Subjective norms	3.24
Environmental citizenship *		
INVOLEMENT IN LOCAL DEMOCRACY	Citizenship	2.38
GOOD COMMUNITY SPIRIT	Citizenship	3.3
RIGHT TO A CLEAN ENVIRONMENT	Citizenship	4.3
Motivations		
INTRINSIC MOTIVE 2	Intrinsic motivation	4.0
INTRINSIC MOTIVE 1	Intrinsic motivation	3.8

* Ascription of responsibility statements also used here
Scales based on a 1 to 5 agreement Likert measure

7 The Exeter Study II: multivariate analyses

Introduction

This chapter focuses on the analysis of the questionnaire data that was collected and described in the previous chapter. The data analysis process outlined in the preceding chapter (Figure 6.1) is followed with varying detail, beginning with a brief account of selected bivariate relationships that exist between the independent and dependent variables within the data set. The questionnaire data are then subjected to a series of multiple regression tests to elucidate the efficacy of each of the independent variables upon the dependent variables. Thus the sequence from data description to high-level multivariate analysis is completed. The chapter ends with a summary of the major empirical findings of the research example.

Bivariate Statistics

The next step after data description in Figure 6.1 leads to bivariate analysis of the data concerned. As stated above, the aim of the Exeter study was to explain, in statistical terms, the behavioural data collected by the questionnaire of Exeter residents. However, as the reader will note from Figure 5.2 there is a further dependent variable that must be explained for the conceptualisation used in the study to function. Behavioural intention is a crucial dependent variable, since this variable is logically a precursor to action and therefore an important variable to explain. More important in building a picture of waste management action, though, are the possible differences in the antecedents of intention and action. This brief section on the bivariate relationships within the data set focus solely on the centre of the conceptual framework in order to show the extent to which the 'logical' process functions.

Behavioural Intention and Behaviour

A crucial part of the conceptual framework used in this study and the central element of the Theory of Reasoned Action is the link between stated intention to act and behaviour. In this study, of course, 'intention' was measured by an individual's 'willingness to act', whilst 'behaviour' was measured by 'reported behaviour'. Despite the obvious methodological problems in separating these two verbal reports, the data in Chapter Six showed that there was the expected difference between 'willingness to act' and 'reported behaviour'. For the framework to have any utility, it had to be shown that there was a statistical link between a stated intention to act and behaviour. To test this relationship, Spearman's Rank correlation coefficient was used. This is a non-parametric test, designed to examine the association between data that are not on continuous scales and do not yield normal distributions of data plots. The degree of association is measured according to whether there is a perfect positive relationship between the two variables (+1) or a perfect negative relationship (-1). A figure of 0 implies no association. The questionnaire data were analysed first according to the total scores for intention related to the total for behaviour. Following this analysis, the total scores for minimisation, reuse and recycling intentions were compared to their behavioural equivalents.

The Spearman Rank correlation for the total of all intention items and all behaviour items was 0.61, significant at 0.05. This implies that there is a moderate to high association between what people state they will do and what they actually do. The significance test given after the test statistic is of great importance. When using tests such as Spearman's Rank, it is necessary to choose a level of significance at which one will accept or reject the statistic. Throughout the research, a level of 0.05 was selected. This implies that there is only a 5% 'probability' that the statistic calculated is due to 'chance' (or 'error'). Thus, a significance level of 0.05 means that one can be 95% 'certain' that the statistic represents a 'true' relationship. In this case, the figure of 0.61 shows that there is certainly a compelling association between willingness to minimise, reuse and recycle waste and what they report they do. This shows the utility of the conceptual framework in positing the relationship between intention and behaviour.

The association between the totals for minimisation, reuse and recycling intentions and behaviour were also calculated using Spearman Rank correlation and were 0.445 for minimisation, 0.474 for reuse and 0.590 for recycling.

All of these statistics were significant at 0.05. They show that, with some variation, there is a moderate to strong link between all three waste management behaviours in terms of the degree to which people are willing to undertake them and the extent to which they do. However, this is where the conceptual framework provides the most important advance in understanding for social scientists. The TRA posited that the only significant predictor of behaviour was intention. As previous studies have shown, in some cases this might be the case, but in the majority, it is not. The statistics shown here provide evidence that there is a logical link from intention to behaviour. At the same time, though, this link is at best moderately strong. Clearly all the variation in behaviour is not explained by that of intention. Thus, behaviour must be 'explained' in other ways other than by intention, as hypothesised in the conceptual framework. The impact of other variables is examined below in the section on multivariate statistics. Suffice to state here that the bivariate statistics have served two key purposes – to support the basic and underlying logic of the intention-behaviour link and to support the conceptualisation of behaviour overall.

Environmental Values, Intentions and Behaviour

The second relationship implied in the centre of the conceptual framework was the logical link between underlying environmental values and behavioural intention. It was argued that in a rational world, there would be strong links between fundamental predispositions towards the environment and the extent to which individuals were willing to act. Table 7.1 shows the Spearman Rank correlation coefficients for each environmental value item in the questionnaire (the wording of which is given in Figure 6.4 above) related to the total score for behavioural intention, as well as the three aggregate scores of the minimisation, reuse and recycling items.

Table 7.1   **Bivariate relationships between environmental values and behavioural intention constructs**

Statement	Behavioural Intention Total	Minimisation Total	Reuse Total	Recycling Total
Strong Sustainability 1	.301*	.421*	.372*	.136*
Strong Sustainability 2	.315*	.361*	.363*	.168*
Ecocentric 1	.296*	.395*	.319*	.131*
Ecocentric 2	.219*	.341*	.297*	.089*
Weak Sustainability 1	.266*	.287*	.294*	.133*
Weak Sustainability 2	.197*	.240*	.255*	.095*
Apathy 1	.422*	.434*	.418*	.252*
Apathy 2	.278*	.362*	.343*	.126*

* Significant at 0.05

The first point to note when examining these statistics is that all the environmental value statements were recoded for ease of analysis. This means that for negatively worded statements, the scores have been reversed so as to make interpretation of the statistics simpler. Thus in the table above, the 'Weak Sustainability' and 'Apathy' statements were reverse coded so that an increase in their score implies, for example, a less apathetic attitude.

Looking at the relationships between environmental values and the total behavioural intention scores in the first instance, it can be seen that there is wide variation according to the value domain being considered. However, none of the statistics is better than moderate, implying that only a weak to moderate positive relationship exists between positive environmental values and a willingness to undertake waste management activities. Thus, the evidence of a logical link within the conceptual framework gains less support from these data. However, it does justify the structure of the conceptual framework where other variables are posited to affect behavioural intention.

This somewhat pessimistic evaluation of the impact of environmental values is lessened somewhat when the individual behavioural elements are examined. It appears that there is significant variation in the statistical associations between the environmental value items and minimisation, reuse and recycling intentions. Moderate correlations exist between all environmental value items and minimisation and reuse intentions, with, in contrast, virtually no association between these items and recycling intentions. Such a finding implies that there is evidence of differential association and, therefore, alternative structuring, of the different behaviours. This supports the data analysis programme set out in Figure 6.1 above where bivariate analysis is followed by analysis of the individual variables to examine whether there are what may be termed 'empirical dimensions' to, for example, behaviour or intention items. From the evidence of the previous chapter and from the associations between environmental values and behavioural intentions, it seems highly likely that for different types of behaviour and intention there are alternative antecedents.

Conclusion

This brief examination of the central element of the conceptual framework has shown how bivariate statistics can be used to examine salient elements of questionnaire data. Evidently, both in this and other research projects, the researcher would progress by examining the bivariate relationships between the external variables in the conceptual framework and those at its centre. There is not space here to examine these relationships, but from this short examination of the central element of the framework, it can be seen that:

- The impelling logic of the framework – the relationship between intention and behaviour – functions moderately well;
- The rationale for implying intervening variables between intention and behaviour is supported;
- The link between environmental values and intention is partially supported, implying different antecedents of intention and, by implication, behaviour; and
- The rationale for suggesting alternative predictors for behavioural intention is supported.

Having established these facts, it was necessary to examine how the conceptual framework functioned at being able to provide a holistic framework of behaviour.

Multivariate Statistics

This section now deals with the nature of the relationships between the situational and psychological factors and intentions and behaviour, as well as the relative influence of each of these within the conceptual framework (see Figure 5.2). The requirement now is to take a broader view and consider the variables not as individual items, but as aggregates that can be used to elucidate multivariate relationships between groups of similar items. This is a very common method when dealing with large data sets in social psychology and geography, and the method used here is similar to that of Derksen and Gartell (1993), Guagnano *et al.* (1995) and Oskamp *et al.* (1991) among others.

Item Interpretation and Data Preparation

Before examining the determinants of behaviour, there are two key processes that need to be undertaken and understood. The data must be prepared for multiple regression analyses such that they are a form suitable for these analyses. This is dealt with below. More importantly though, they must be understood in terms of their 'empirical dimensions'. It was shown above that environmental values had a moderate association with intentions to reuse and minimise waste, but very weak associations with recycling intentions. There has, thus far, been the logical assumption that questionnaire items defined by the researcher function best within the groups defined by that researcher. Yet empirically this may not be the case. For example, the researcher in this case defined four fundamental environmental value groupings. In statistical terms, these may not actually have much relationship with each other since individuals filling out the survey may have interpreted them differently. Similarly, with the psychological factors, there may be alternatives to the groupings provided by previous authors and as shown in the conceptual framework. Analysing the items so that they can be grouped according to empirical responses allows the researcher to examine both the way that individuals have answered the survey and what new concepts might be more appropriate for conceptualising behaviours, intentions, values and psychological factors.

Social scientists have developed a useful technique for examining the 'empirical dimensions' of questionnaire data measured on ordinal scales, such as the Exeter Survey. It also provides data in a format that can be used for the explanatory statistics to be examined below. The technique has become known as 'factor analysis', but can also be termed 'principal components analysis'. Essentially, the technique examines questionnaire items by assessing the degree to which each item correlates with a series of hypothetical 'factors'. By a series of statistical procedures, the computer programme will define a number of 'factors' which contain the items from the questionnaire. The items are said to have 'loaded' onto these factors because they have the best association with them. The factors containing items may look very different from what the researcher might expect.

In the Exeter study, the behaviour, behavioural intention, environmental value and psychological variables were 'factor analysed'. It is helpful to examine where the questionnaire items 'loaded' so as to both understand the data and to see the differences between what emerges compared to what was expected.

Table 7.2, as an example, shows the 'factor loadings' for the behaviour items. The bold figures identify the individual factors. These are correlation coefficients and simply show how much association there is between each item and the respective factor. The '% Var' row is an expression of how much each factor explains the total variance in the data (i.e. .346 = 34.6%). This table acts to demonstrate the technique, but what is of importance is the three 'factors' that emerge. As can be seen, they conform almost precisely to what was expected. Factor 1 contains all ten of the recycling items, whilst the other two factors contain the minimisation and reuse items. Clearly, individual behaviour is focused around these three fundamental actions, as conceptualised and now shown by how individuals have answered the questionnaire.

Table 7.2   'Factor loadings' for all behaviour items

Variable	Factor 1	Factor 2	Factor 3
Recycle cardboard	**0.899**	-0.093	0.042
Recycle junk mail	**0.896**	-0.076	0.049
Recycle plastic bottles	**0.895**	-0.055	0.062
Recycle magazines	**0.867**	-0.092	0.152
Recycle food cans	**0.837**	-0.192	0.045
Recycle drinks cans	**0.826**	-0.088	0.125
Recycle foil	**0.825**	-0.139	0.088
Recycle newspaper	**0.805**	-0.115	0.204
Recycle textiles	**0.705**	-0.289	0.076
Recycle glass	**0.596**	-0.295	0.268
Loose produce	0.080	**-0.789**	-0.036
Less packaging	0.187	**-0.779**	0.135
Own bag	0.116	**-0.700**	0.182
Buy re-usable	0.059	**-0.667**	0.320
Buy recycled	0.213	**-0.641**	0.253
Wash dishcloths	0.148	**-0.584**	0.133
Reuse bottles	0.127	-0.202	**0.813**
Reuse paper	0.163	-0.136	**0.812**
Reuse tubs	0.148	-0.262	**0.696**
Repair items	0.031	-0.482	**0.498**
% Var	0.346	0.178	0.123

Bold figures show the individual factors

For the behavioural intention items, two factors emerged which conformed exactly to one 'recycling factor' and another 'minimisation/reuse factor'. This is of interest since as the section on bivariate statistics showed, there was a significant association between minimisation and reuse and environmental values, but not between recycling and environmental values. The data from Figure 6.3 also show less difference between minimisation and reuse intentions than for behaviour. It therefore appears that individuals show more divergence between behaviours than intentions.

Of considerably more interest is the analysis of the environmental value and psychological items. Table 7.3 shows the way the environmental value items 'loaded'. Unlike behaviour and behavioural intention, the items have loaded on fewer and somewhat unexpected factors. This shows perfectly the value of factor analysing the variables in question. Factors 1 and 2 are at first inspection difficult to separate. Yet Factor 1 may be seen as displaying traits of the previously mentioned 'strong sustainability' concept (remembering that the 'Apathy' and 'Weak Sustainability' items have been recoded). The emphasis placed upon human elements of environmental protection is noticeable. Yet this could not be said to be in any way 'technocentric' in the sense of Timothy O'Riordan's (1985) meaning. Rather, it reflects some kind of 'human priority' within the environmental context. Factor 2 is simpler to assign a label to. This is more tilted towards an ecocentric or 'importance of nature' viewpoint.

Both of these factors demonstrate that individuals held positive values concerning the environment, varying in the extent to which humans were significant. This has already been shown by Figure 6.4. However, of additional interest is item 'Ecocentric 1', which 'loads' heavily on both factors. In this case, the item is excluded, as it is not clear where it can be placed.

Table 7.3   'Factor loadings' for all environmental value items

Variable	Factor 1	Factor 2
Apathy 1	**0.835**	0.129
Weak sustainability 1	**0.821**	0.037
Strong sustainability 2	**0.690**	0.329
Strong sustainability 1	**0.643**	0.388
Ecocentric 1	0.570	0.508
Ecocentric 2	0.062	**0.842**
Weak sustainability 2	0.196	**0.762**
Apathy 2	0.464	**0.553**
Variance	2.8438	2.1309
% Var	0.3355	0.266

Apathy 1: The environment is of little concern to me
Weak sustainability 1: Nature isn't harmed by human changes
Strong sustainability 2: If we over-use our natural resources, human development may be harmed in the future
Strong sustainability 1: The environment is forgotten too often when decisions are made
Ecocentric 1: Nature and the environment have as much value as human beings
Ecocentric 2: Humans should not develop any more resources or land, in order to protect the natural environment
Weak sustainability 2: Human welfare should be our primary concern in the future
Apathy 2: Getting through daily life and surviving is what concerns me the most, not the environment
Bold figures show the individual factors

Overall, the analysis of the environmental value items has shown that only two environmental value constructs are of importance in the Exeter study, and not the four previously tested. Such analysis provides context for the later findings and enables insight into how individuals value their natural surroundings.

Table 7.4 'Factor loadings' for all psychological items

Variable	1	2	3	4	5	6	7	8	9
STORAGE SPACE 2	**0.829**	0.150	0.005	0.127	0.041	-0.026	-0.039	0.110	-0.008
RECYCLING CONVENIENCE 2	**0.825**	0.128	0.080	0.102	0.035	0.011	-0.234	0.011	-0.088
STORAGE SPACE 1	**0.824**	0.078	0.084	0.132	0.010	-0.110	0.012	0.178	0.016
RECYCLING CONVENIENCE 1	**0.806**	0.061	0.128	-0.031	0.032	-0.097	-0.131	0.078	-0.118
SELF EFFICACY 2	**0.667**	0.309	0.093	-0.191	0.081	-0.158	-0.202	0.090	-0.099
SELF EFFICACY 1	**0.537**	0.363	0.037	0.125	0.032	0.065	-0.329	0.061	0.042
TIME 1	0.246	**0.740**	0.122	0.235	0.147	-0.052	-0.018	0.130	-0.083
TIME 2	0.298	**0.717**	0.138	0.069	0.125	-0.037	0.118	0.154	0.093
ASCRIPTION OF RESPONSIBILITY 2	0.125	**0.677**	0.155	0.244	0.187	-0.033	-0.102	-0.155	-0.177
WASTE CONCERN 1	0.164	**0.645**	0.61	0.296	0.211	-0.053	-0.232	-0.033	-0.154
OBLIGATION TO ACT 1	0.086	**0.611**	0.255	0.368	0.105	0.028	-0.175	-0.021	-0.100
WASTE CONCERN 2	0.122	**0.606**	0.264	0.127	0.249	-0.289	0.037	0.098	-0.212
OBLIGATION TO ACT 2	0.113	**0.513**	0.332	0.289	0.067	-0.325	0.137	-0.014	-0.026
BELIEFS ABOUT WASTE MANAGEMENT 3	0.103	0.090	**0.802**	0.128	0.117	-0.111	-0.014	0.075	-0.119
BELIEFS ABOUT WASTE MANAGEMENT 2	0.090	0.163	**0.793**	0.132	0.132	0.002	0.038	-0.117	-0.102
BELIEFS ABOUT WASTE MANAGEMENT 4	0.068	0.349	**0.617**	0.038	0.110	-0.141	-0.047	0.073	-0.063
BELIEFS ABOUT WASTE MANAGEMENT 1	0.051	0.185	**0.548**	0.237	0.151	0.056	-0.196	-0.035	-0.332
RESPONSE EFFICACY 2	0.013	0.303	0.092	**0.776**	0.072	0.031	-0.085	0.105	-0.013
INTRINSIC MOTIVE 2	0.122	0.310	0.159	**0.696**	0.117	0.013	0.049	-0.185	0.075
RESPONSE EFFICACY 1	0.125	0.333	0.213	**0.603**	0.183	-0.175	0.018	0.117	-0.266
INTRINSIC MOTIVE 1	0.051	0.100	0.230	**0.462**	0.122	-0.365	-0.044	0.069	-0.288
THREAT 1	0.018	0.195	0.102	-0.003	**0.857**	-0.112	0.058	0.202	-0.106
THREAT 2	0.069	0.330	0.102	0.150	**0.831**	0.034	-0.094	0.005	0.112
WASTE PROBLEM PERCEPTION 2	0.038	0.268	0.255	0.268	**0.564**	-0.131	-0.036	-0.182	-0.138
WASTE PROBLEM PERCEPTION 1	0.042	-0.021	0.198	0.099	**0.528**	-0.206	0.081	-0.011	-0.277
ACCEPTANCE NORM TO RECYCLE 2	0.144	0.232	0.147	0.062	0.073	**-0.814**	-0.045	0.061	0.060
ACCEPTANCE NORM TO RECYCLE 1	0.052	0.060	-0.034	-0.024	0.118	**-0.740**	-0.269	0.033	-0.045
AWARENESS NORM TO RECYCLE 2	0.379	0.021	0.033	0.086	-0.59	-0.129	**-0.759**	0.129	0.077
AWARENESS NORM TO RECYCLE 1	0.380	0.079	0.052	-0.043	0.044	-0.233	**-0.730**	0.164	-0.031
INVOLEMENT IN LOCAL DEMOCRACY	0.182	0.006	-0.075	-0.074	0.038	-0.079	-0.022	**0.831**	-0.110
GOOD COMMUNITY SPIRIT	0.276	0.029	0.097	0.134	0.079	-0.026	-0.277	**0.738**	0.029
RIGHT TO LIVE IN A CLEAN ENVIRONMENT	0.046	0.359	0.266	-0.045	0.0912	0.059	0.044	0.107	**-0.763**
ASCRIPTION OF RESPONSIBILITY 1	0.161	0.127	0.106	0.245	0.206	-0.082	0.003	-0.031	**-0.632**
% Var	0.129	0.127	0.081	0.076	0.074	0.055	0.051	0.048	0.048

Exact statement wording can be obtained from the tables in Section 6.2 above
Bold figures identify the factors extracted

Finally, Table 7.4 shows the factor loadings for all of the psychological items in the questionnaire. The change from the ordering within the conceptual framework (described in Table 6.1) is dramatic. Yet examination of the factors on the whole makes sense. Factor 1 is almost solely concerned with 'efficacy' factors, such as convenience of recycling, storage space for materials and, of course, 'self efficacy'. Such a factor essentially comprises 'convenience and effort' factors.

Factor 2 is a mix of statements which, although diverse, can be termed 'active concern' since there are statements examining elements of responsibility towards the environment, the concern felt about the waste problem and moral obligation to act appropriately.

Factor 3 contains all four 'belief' items, representing a factor that can be termed simply 'beliefs about waste management'. The combination of 'response efficacy' and 'intrinsic motivation' provides an interesting fourth factor. The combination of rationales to act (i.e. a belief that action has a tangible effect) along with an intrinsic satisfaction from acting, provides a new factor which can be termed 'motivation to respond'. This scenario repeats itself with factor 5, where 'threat' and 'waste problem perception' emerge in a single factor. It appears that those who perceive a problem are likely to also be those who feel most threatened by the waste problem, resulting in a 'problem/threat' factor.

Factors 6 and 7 are logical, containing the acceptance and awareness of the norm to act, as anticipated. Finally, the 'citizenship' factors emerge. First as factor 8, showing the items concerned with local democracy and community ('community/democracy') and second factor 9 which has the classic 'rights and responsibilities' combination, earning it the title of 'citizenship'.

Thus, these 'factor analyses' have shown how differently many of the factors from the conceptual framework have 'loaded' and hence how it is unwise to proceed with one notion of how items are linked. Clearly, the residents of Exeter have different ways of, for example, looking at how they perceive the waste problem. From this analysis, it can be seen that the following factors emerge:

- 3 behaviour factors: 'minimisation', 'reuse' and 'recycling';
- 2 behavioural intention factors: 'minimisation/reuse' and 'recycling';
- 2 environmental value factors: 'human priority' and 'importance of nature'; and

- 9 psychological factors: 'convenience and effort', 'active concern', 'beliefs about waste management', 'motivation to respond', 'problem/threat', 'acceptance of norm to recycle', 'awareness of norm to recycle', 'community/democracy' and 'citizenship'.

As was mentioned earlier, such factors can be used to formulate 'scales' for multivariate analyses. Essentially, by summing the items of each scale, an acceptable interval scale, which is normally distributed, can be formed for such analyses.

Multiple Regression Analyses

In order to find out which factors influence behaviour the most, all of the data from the questionnaire were analysed using a technique called 'multiple regression'. This advanced method of correlation analysis assesses the effect of each independent (or predictor) variable on the dependent variable. To first find out which factors are significant in changing the dependent variable, all of the independent variables are analysed by what is known as 'stepwise' regression where the computer progressively eliminates independent variables that have no impact on the dependent variable. The final sub-set of variables is then 'regressed' using normal multiple regression and the computer calculates how much effect each of the independent variables has on the dependent variable.

In the case of the Exeter data, there were five 'dependent' variables which were identified by the factor analyses above. These are shown on the top row of Table 7.5 and relate to behaviour and behavioural intention. The 'independent variables' are given in the left-hand column of Table 7.5 and relate to:

- behavioural intention (which is used to 'predict' behaviour);
- environmental values ('Human Priority' and 'Importance of Nature' scales);
- psychological factors ('Convenience and Effort' scale down to 'Citizenship' scale);
- socio-demographics (age, gender, house type);
- community group membership ('total group membership');

- knowledge ('environmental', 'local waste' and 'policy', plus 'knowledge sources', a measure of information sources used about waste issues);
- 'experience';
- 'recycling provision index' (index of the number of items recycled at respondents nearest static site); and
- 'kerbside collection bin' (access to 'Recycle from Home' scheme).

Table 7.5   Final beta weights for dependent variable scales

Independent variables	Intention Minimise-Reuse	Intention Recycle	Behaviour Minimise	Behaviour Reuse	Behaviour Recycling
Intention Minimise-Reuse			0.29	0.27	
Intention Recycle					0.33
Human Priority	0.13		0.13		
Importance of Nature	0.11	-0.06	0.11	0.1	0.05
Convenience and Effort		0.2		0.15	0.2
Active Concern	0.25	0.2			
Problem and Threat	0.17				
Motivation to Respond				0.2	
Awareness of norm					0.1
Acceptance of norm	0.07	0.24		-0.08	-0.07
Community/Democracy			0.13		
Citizenship	0.11				
Age			0.16		0.07
Gender	0.07		0.11		
House Type		0.1			
Total Group Membership				0.11	
Environmental Knowledge	0.07				
Local Waste Knowledge		0.21			0.19
Policy Knowledge			0.11		
Knowledge Sources		0.08	0.15		
Experience	0.16			0.04	
Recycling Provision Index		-0.08			
Kerbside Collection Bin	-0.2				0.28
R^2_{adj} (per cent variance explained)	50%	53%	43%	31%	79%
F (significance, all <0.05)	63.5	87.7	60.0	41.2	293.5

Table 7.5 only shows the effect of an independent variable if that variable had a significant effect on the dependent variable in the light of all others. Hence, the independent variables in Table 7.5 are the 'best predictors' of the respective dependent variables. Each figure is a 'final beta weight', which means that they are statistics which have been sorted by stepwise regression and that each statistic alludes to how much the

dependent variable will change when a change in the independent variable is seen. Standardised coefficients enable changes in one variable to be compared to all others, since the measurement scale is the same (or 'standardised'). In essence, the coefficient tells the researcher how many standard deviation points the dependent variable will change (in both positive and negative terms) for a one-unit move in the independent variable. This need not concern the reader any further; the key point to bear in mind is that the greater the statistic, the more change is seen in the dependent variable and the greater the impact of the independent variable. At the bottom of each column (representing each analysis) R^2_{adj} and F statistics are shown. The former gives an estimate of the amount of variance in the dependent variable is explained by the independent variables, expressed as a percentage and adjusted to take account of the number of independent variables in the model. The latter statistic is an expression of the overall significance of the model (Fisher's F ratio), derived from the degrees of freedom in the model (the number of repressors and cases). Again, all one need bear in mind is that the larger the statistic, the more significant the model.

In addition to Table 7.5, Figures 7.1 to 7.4 provide the information in Table 7.5 in the form of a path analysis. This enables more simple analysis of the Table and also provides more detailed information concerning the efficacy of individual factors and their impact upon the dependent variables. These diagrams are designed such that they can be interpreted in the same manner as Figure 5.2 (the conceptual framework). Thus, the logical processual movement from values to intention (or willingness to do x to behaviour x) is in the centre, with the situational variables at the top and the psychological factors on the bottom rows. Each path coefficient (standardised regression coefficient) is given along with an arrow that denotes its importance by the thickness of the line. Additionally, error terms (unique to path analysis) are given to assess the proportion of movement in the dependent variable not explained by the independent variables. For those variables that have an impact upon both willingness to do x and behaviour x the overall impact of that variable upon behaviour (here the end point and most crucial of the two dependent variables) is shown in italics by that variable label. This is derived from the multiplication of indirect path coefficients (to gain the variable's indirect effect on behaviour) and then added to its direct effect on behaviour. For variables that only have an indirect effect on behaviour (via willingness to do x) this effect has been calculated and given in italics adjacent to the variable label.

Thus path analysis, whilst in this case merely being a representation of the regression table, gives a clearer graphical account of the data.

Figure 7.1   Path diagram of minimisation behaviour showing the principal factors that predict a willingness to minimise and actual action

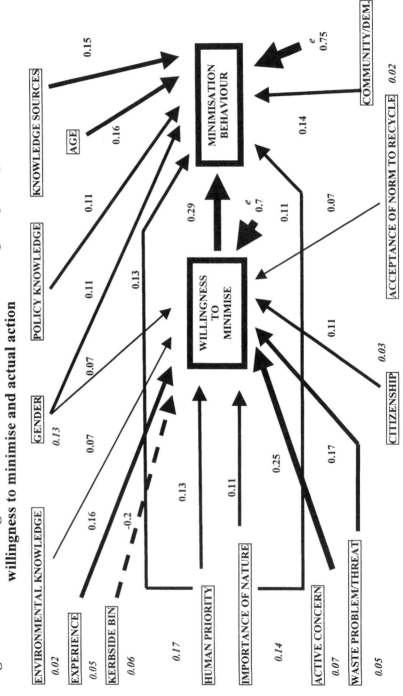

Figure 7.2 Path diagram of reuse behaviour showing the principal factors that predict a willingness to reuse items and actual action

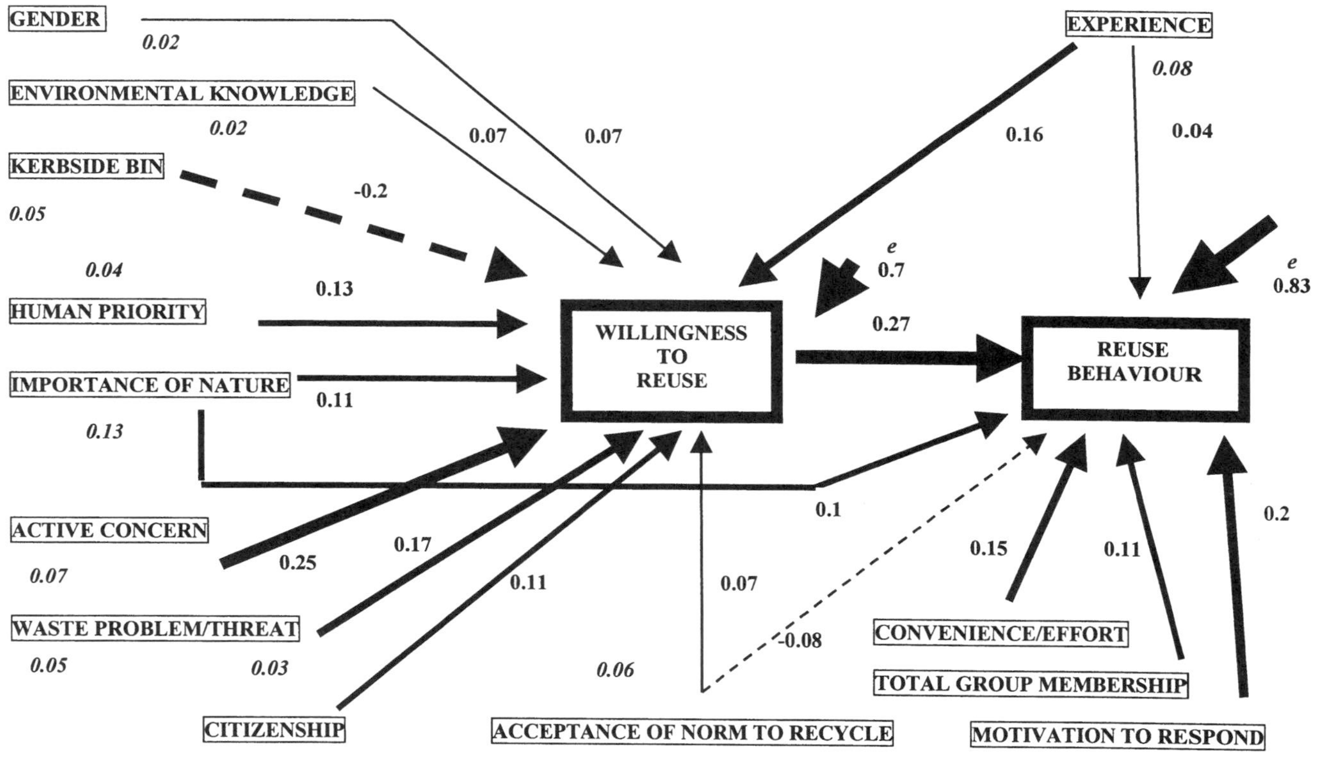

Figure 7.3   **Path diagram of recycling behaviour showing the principal factors that predict a willingness to recycle and actual action**

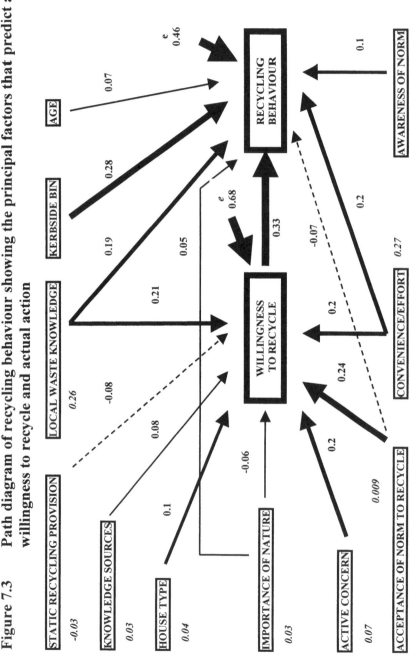

Analysis of Results

The three path diagrams represent the three types of behaviour identified by the factor analysis above. However, as the reader will have noted, the willingness to act scales are only divided into two (logical) factors. Thus, for both minimisation and reuse behaviour the theoretical assumption is that the attitudinal underpinnings are the same, whereas in reality they are probably slightly different. However, reference to Table 7.5 shows that the relationship between these factor scales and both sets of behaviours is both very significant and similar.

Minimisation   In terms of the efficacy of the models, Table 7.5 shows that the R^2_{adj} and F statistics for intentions towards minimisation/reuse and minimising behaviour. Neither fit is very good, but the significance is acceptable in both cases. Figure 7.1 shows that in the cases of both intention and behaviour the error terms are high and that there is considerable unexplained variance in these dependent variables. Nevertheless, as will be seen below, the conclusions from these models are logical and significant, even if the fits are rather weak.

Reference to the first and fourth columns in Table 7.5 and Figure 7.1 immediately shows a situation similar to that envisaged in the theoretical conceptual framework in Figure 5.2. Remembering that the variables shown are only those variables that were significant enough in the framework to be retained in stepwise analysis, it can be seen that, as anticipated, there are indeed variables that influence intention and behaviour independently. There are indeed a number of intervening variables in the relationship between intention and behaviour that have no significant impact upon intentions themselves (in this case age is a good example). Most crucial, however, is that the strongest relationship is between willingness to act and actual behaviour. Thus, the evidence given here supports the theoretical link between intention and behaviour.

Intention   If intention is examined first, as the logical precursor to behaviour, it can be seen that holders of both sets of environmental values (Human Priority when protecting nature and Importance of Nature) have almost equal impact on movement of the dependent variable. It was hypothesised above that minimisation (as well as reuse, see below) behaviour is less ingrained into daily life than recycling behaviour and therefore underlying (positive) environmental values may play a more prominent part in the prediction of pro-reduction intentions and behaviour. Certainly here, although the impact is not huge, it is significantly higher than the predictive value for recycling (see below). It would suggest, quite rationally, that those who hold positive views about the importance of the natural world and the need to protect nature for future generations are more likely to be willing to reduce their waste throughput by minimising their production of such wastes. This finding gives tacit support both to the work of writers such as Steel (1996) of recent years, and Weigel and Weigel (1978) and Dunlap and Van Liere (1978) of the pioneer years that underlying environmental values are important.

Looking at the situational characteristics that impact upon minimisation intentions, the most important predictor appears to be presence or absence of a recycling bin at the kerbside. In this case it is seen that the presence of a bin is likely to lead to a significant reduction in one's willingness to minimise. Again, this makes logical sense. If the above hypothesis concerning the internalisation of norms regarding specific behaviours is correct, then it could be stated that those with a recycling bin may feel that because they 'do their bit' each week by recycling their waste as a matter of course, they are unlikely to be aware of the more socially marginalised intentions and behaviours regarding reducing and reusing wastes. A further explanation could be that of De Young (1986) who asserts that making waste management too easy for people is likely to lead to an overall reduction in waste behaviour amongst some individuals since they no longer feel intrinsically motivated to undertake the behaviour. This is because the external circumstances have become such that it is no longer a behaviour that requires effort. Nevertheless, it is an important finding that access to kerbside recycling, whilst boosting recycling rates, is likely to lead to a reduced willingness to minimise wastes.

Conversely, the next most significant situational factor predicting willingness to reduce waste is experience of recycling. This is apparently an anomaly, since those who have experience are also exclusively those with a recycling bin. However, the reader will note that those with experience are comparatively small, and by no means represent the kerbside recycling community as a whole. Again, the finding makes logical sense. Those with experience are the people who recycled before having a kerbside collection. They are therefore more likely to be those who are willing to act in ways that are not situationally simple or socially 'normal'. Hence, with reference to the previous finding, it seems practical to suggest that the data point towards an overall negative effect upon minimisation intentions from those who have a kerbside recycling bin, but that this relationship is reversed when experience of recyclers is accounted for. This again is an important finding since it supports the idea that a willingness to reduce waste is more marginal than, say, recycling.

Finally, the two other situational factors that have a significant impact in the regression framework are environmental knowledge and gender. Their impact is minor as compared to the other factors. However, it is not odd to suggest that some increase in global environmental knowledge might lead to a more positive intention to minimising wastes. The slight impact of gender is interesting, but compared to its impact on behaviour is minor (see below).

In terms of the psychological variables in the framework, the most prevalent is active concern about waste. This would appear to be one of the most important factors, in theory at least, that should shape minimisation intentions. The reader will recall that this scale is made up of items that emphasise concern over waste issues, the guilt at not acting responsibly and the relation of waste issues to oneself. Clearly, the more concerned one is and the guiltier one feels when not acting, the more likely it is this guilt will impact on a willingness to reduce waste.

Similarly, the more of a problem that waste appears to be (the next most important psychological factor) the more willing people are to minimise. This may have more of an impact than on recycling intentions because those who understand the problem more, may realise that it is both waste production, as well as how waste is dealt with, that is a core issue.

Those who believe that they have the right to live in a clean environment and that it is everyone's responsibility to ensure that this comes about, are more likely to be willing to reduce their waste. Under the umbrella term 'citizenship', this factor emphasises the individual element in environmental responsibility. It is therefore not surprising that it has a significant impact on a willingness to reduce waste since this is something that is, on the whole, only achievable by the individual (like reuse), whereas the recycling of waste is usually perceived as being an industrial process undertaken by the state.

Finally, there is a weak predictive value from the acceptance of the norm to recycle. Although not immediately obvious why this variable is included (apart from correlation by chance) it may be that this narrow statement (referring only to recycling) also has some merit with regard to minimisation. It could be that those who are more ready to accept those around them recycling may also be those who are willing to accept other norms. However, this is unsupported and a very tentative suggestion.

Behaviour   It is obvious from Figure 7.1 that minimisation behaviour is related strongly to a willingness to reduce waste. Hence, the amalgamation of the previously discussed factors into an 'intention' concerning reducing waste will indirectly affect behaviour. However, it can be seen that there are a number of intervening variables that have a significant direct effect upon minimisation behaviour.

The two environmental values items (Human Priority when protecting nature and Importance of Nature) again share almost equal amounts in their impact upon behaviour which are interestingly very similar to their impact on intention. Clearly there is variance in minimising behaviour that is explained better by these factors than by intention, even though they impact upon this as well.

The most significant situational factor involved in shaping minimising behaviour independently of intention is age. It would therefore appear that despite any good intention (or not) to reduce waste, those who are older tend to minimise more anyway. This finding is consistent with that of Hallin (1995) who discussed this phenomenon in the context of American age cohorts, relating enhanced recycling behaviour to the Second World War generation. This finding also dismisses a core element of early 1980s American research that asserted the link between youth and environmental behaviour (see Van Liere and Dunlap, 1980, for a review of the literature).

Of the situational factors involved, the second most significant of these is where people obtain knowledge. It is seen that the more sources of knowledge people have available to them, the more they tend to minimise, even though this is apparently insignificant at shaping a willingness to minimise. Logically, it could be argued that even though a person might be willing to minimise, a lack of knowledge of how to do this (derived from few sources of knowledge) could hamper their good intentions. A wider knowledge base therefore appears important in transferring aspirations into action.

Both policy knowledge and gender have the weakest effects on behaviour of the situational variables. It appears that knowledge of specific policy issues (Local Agenda 21 and Sustainable Development) is likely to intervene to positively influence behaviour. This is probably related to knowledge sources in the sense that increased use of different media to obtain knowledge leads to enhanced policy knowledge. In terms of gender, women are more likely to minimise. This may not be surprising if it is seen that although men may have positive intentions toward reducing waste, they are not actually the ones who do it. It is therefore likely to depend on household lifestyles more than a crucial difference between intentions between men and women regarding reducing waste, as Figure 7.1 shows.

Finally, the one psychological factor that intervenes in the intention-behaviour relationship is the Community/Democracy factor, which is made up of statements examining personal perception of community well being and inclusion in local decisions on the environment. It is likely that this factor is more of a correlative anomaly than representing a true relationship. However, there are two possible reasons for its status. First, it could be that those who hold such views are more involved in the community and by default are involved in environmental schemes, such as community composting or Local Agenda 21. On the other hand, they may feel happier with the state of local government and feel more willing to participate in local efforts to reduce waste. These suggestions are, however, tentative.

Overall summary Considering the above and the overall effects in Figure 7.1, minimisation behaviour can be predicted thus:

• Willingness to minimise/reuse (predicted by environmental values, active concern, kerbside bin, waste problem/threat, experience and citizenship).

And:

• Age;

• Knowledge sources;

• Community/Democracy scale;

• Human priority when protecting nature scale (overall effect = 0.17);

• Gender (overall effect = 0.13);

• Importance of nature scale (overall effect 0.14);

• Policy knowledge.

This large number of factors with moderate regression coefficients suggests a complex behaviour pattern, especially when the overall explanatory power of the model is seen to be only 43%.

Reuse Table 7.5 shows that whilst the intention scale has a reasonably good fit (50%, as above for minimisation), the reuse behaviour scale is a poor fit (31%). Nevertheless, the model is significant. Figure 7.2 demonstrates this situation in the form of a path analysis and shows that the error term for reuse behaviour is indeed high.

Although the left hand section of Figure 7.2 is slightly different from Figure 7.1 (minimisation) it will be noted that all the predictors of the so-called 'willingness to reuse' scale are the same as 'willingness to minimise' since these are one and the same factor scale. However, the interaction of certain variables between intention and behaviour is different, as indeed are the coefficients of overall effect on reuse behaviour. As can be seen, experience of recycling now impacts both on intention and behaviour, as does acceptance of the norm to recycle. However, gender now only impacts on intention.

Since the determinants of intention are the same as above for minimisation, the following section only discusses reuse behaviour (see above for a full interpretation of the factors influencing intention).

Behaviour Again, the principal factor acting on reuse behaviour is willingness to undertake such activities. However, there are other factors. On the situational side, only experience has a small impact directly on reuse behaviour. Given its larger overall impact, and the fact that most of the effect is indirect, there are few conclusions that can be drawn here, except to state that experience of recycling before having a kerbside collection may act in some cases to enhance a willingness to reuse. However, the majority of the path is seen to pass via willingness to reuse.

A very significant direct effect is exerted from the Motivation to Respond variable among the psychological factors. It would appear that a belief in the efficacy of an action to produce a valuable environmental result and enjoying that activity, plays an important role in shaping reuse behaviour. An aspiration to reuse more may therefore be important, but believing in one's response and enjoying it appears to make a large difference too. This is consistent with De Young's (1986) analysis of intrinsic motivation and Kantola *et al.*'s (1982) discussion of efficacy.

A further important factor that influences only behaviour is 'Convenience/Effort'. Although the statements in this scale refer to recycling, they could also refer to reuse. For example, lack of space to store items that could be reused is likely to be one example when such a factor could have a negative effect. In this case, increase in convenience and reduction in effort required appears to influence reuse behaviour.

Total group membership appears to have a direct effect on reuse behaviour. This may relate to the argument made above concerning minimisation behaviour. It could well be that those in such groups are already engaged in such activities as reuse, and so this intervenes directly to modify their behaviour. Alternatively, more involvement in community activities may lead to an enhanced acceptance of new initiatives.

Finally, there is an anomalous relationship between acceptance of norm to recycle and reuse behaviour, which is negative. Reference to Figure 7.2 shows, however, that the overall effect of this variable is positive and thus the negative statistics can be ignored.

Overall summary   Considering the above and the statistics for overall effect given in Figure 7.2, reuse behaviour is predicted most effectively by:

● Willingness to reuse/minimise (predicted by environmental values, active concern, kerbside bin, experience, waste problem/threat and citizenship).

And:

● Motivation to respond (response efficacy and intrinsic motivation);

● Convenience/effort;

● Total group membership;

● Importance of nature (overall effect of 0.13).

Thus, a complex set of variables determines reuse behaviour, but with only 31% of this variance explained, this is by no means a perfect or complete model.

Recycling   Table 7.5 gives the R^2_{adj} and F statistics for willingness to recycle and recycling behaviour. Although intention has a reasonably good fit (53%), behaviour is exceptionally well fitted (for human geography, at least) at 79%. The corresponding significance values are acceptable. This situation is shown graphically in Figure 7.4. As can be seen, the error terms are much lower than for minimisation or reuse intentions/behaviour.

Before going on to assess the individual components, it is crucial to note that the determinants of intention, behaviour and their interaction are fundamentally different from reduction and reuse behaviours. Not only are there fewer variables involved, but the nature of the factors is also fundamentally altered.

Intention   The importance placed upon environmental values in the previous two assessments is much reduced when recycling is examined. Figure 7.4 shows that there is a direct negative effect on willingness to recycle (a phenomenon that is reversed when behaviour is assessed, see below). This may reflect what was alluded to above, namely that environmental values play a significant role in shaping intentions towards behaviours that are more marginal socially (such as waste reduction and reuse), but that their role, if any, is minor in shaping well-established recycling intentions. In this case, there is a slight negative effect, meaning that an increase in scores on the importance of nature scale will lead to a slight decrease in willingness to recycle. This is theoretically unlikely, and when the total effect of this variable is assessed, it is seen that its overall effect is very slightly positive (see below). Nevertheless, it is crucial to note the diminishing importance of this variable when recycling is considered.

Of the situational variables that impact on willingness to recycle, local waste knowledge (i.e. knowledge of static and, where appropriate, kerbside facilities for recycling) is the most important in influencing this variable. This is logical, since people are unlikely to be willing to recycle if they cannot think of where, when, how and what to recycle. Hence, it is not surprising that such a variable influences intention. Neither is the second most important variable, house type, a surprise when it comes to assessing willingness to recycle. This is because the storage of recyclables is likely to be more difficult in smaller homes. Storing such items in flats is almost impossible and in terraced houses it is still difficult. Hence, those in larger homes will have more room to keep items for recycling and are likely to be more willing to do so.

The two other factors that have an effect on willingness to recycle are knowledge sources and static recycling provision. The former has a slight positive effect, and this is not surprising since alternative information on any aspect of recycling is likely to enhance one's intention. It is not clear how the increase in kerbside sources directly impacts on intention, but this could be in a number of ways, such as increasing awareness of the need to recycle, or perhaps how to recycle. The second factor, static recycling provision, has a negative impact on recycling intentions. This is very surprising, since it would be expected that more recycling provision would lead to a better intention to recycle. However, this would depend upon people recycling in accordance with the measuring schema used in this study, namely that they use the site nearest to them. Although detailed, this scale cannot take account of those who may use their car to recycle materials, as many probably do. Hence, this rather odd relationship between static provision and a willingness to recycle can thus far be ignored, although as will be seen below, it is a useful predictor of other scales.

In terms of the psychological factors that explain the variance in willingness to recycle, the strongest predictor is acceptance of the norm to recycle. The two statements that make up this scale assert that respondents recycle more when they see others recycling and that the more others around them recycle, the more they will. This appears to have a very significant impact on people's willingness to recycle. The presence of significant others recycling therefore appears to influence personal evaluations about whether to recycle. However, as will be seen below with regard to behaviour, the overall impact of this variable is low, and norm acceptance, whilst transferring into a positive intention, may not relate strongly to anything else.

As with minimisation and reuse, active concern about waste issues is important when considering willingness to recycle (although less pronounced). This weaker predictive value may indicate that being concerned about waste and feeling guilty when not acting correctly, is less important in shaping a willingness to recycle than for other environmental intentions. This may be, again, because recycling is more of an established behaviour and factors such as environmental values and waste concern are less important.

Convenience and effort is a further important factor in predicting a willingness to recycle. Clearly, those who are more willing to recycle are those who find recycling sites convenient, who have space to store recyclables and who find recycling easy. This is likely to affect how willing someone is to recycle (as well as his or her actual behaviour, see below), because those who have to travel a long way to a recycling site, have little room to store tin cans, etc., and most importantly those who feel it's too difficult to recycle, are those most likely to be unwilling to recycle.

Behaviour The very good fit provided by the regression model applied here gives more confidence when discussing the results of this particular analysis. As has occurred before, the 'importance of nature' variable acts to modify behaviour as well as intention. However, this modification is very small, and it is seen that the overall effect of this variable is very small indeed (0.03).

In terms of the situational variables, the presence or absence of a kerbside recycling bin is crucial in shaping recycling behaviour. Although this had no significant effect on willingness to recycle, suggesting that presence of a kerbside bin in general has little impact on intention, in practice, recycling behaviour is shaped significantly by this important factor. It is seen that this variable is not that much less significant in modifying recycling behaviour than intention. Hence, presence or absence of a bin is absolutely crucial in improving recycling behaviour. Of less importance to behaviour alone, but of more significance overall, is local waste knowledge. Its direct effect is 0.19, but overall effect is 0.26. This shows that knowledge of what can be recycled where is crucial to recycling behaviour, as has been suggested and would be expected.

Finally for the situational factors, age has a slight positive direct effect on recycling behaviour. This is not too surprising, but the relationship is quite weak and it could not be said that being older was a good predictor of recycling behaviour by any means, as much of the North American literature would assert (see Van Liere and Dunlap, 1980, and Steel, 1996).

With regard to the psychological variables, a significant amount of variance is explained by the 'Convenience/Effort' factor that also had the same level of impact on willingness to recycle. It is logical that this factor would also intervene to shape recycling behaviour as well as intention, since even those willing to recycle might still perceive difficulties when actually undertaking the action is considered. As with local waste knowledge, the overall effect of this variable on recycling behaviour is more than either the indirect or direct effects, such that the overall effect is 0.27. As was expected, the convenience of recycling, storage space available and perceived difficulties are all crucial when analysing behaviour.

Another psychological factor that impacts on recycling behaviour is the awareness of the norm to recycle. It was expected that this factor would only operate indirectly (through acceptance of norm) to modify recycling behaviour. However, it appears that the statements referring to awareness of others recycling is an important variable in itself, and may in fact be a better measure of the 'acceptance of norm' than the measure used here, since *accepting* that one acts because others do is not a common trait in society. It therefore appears that awareness of others acting does have some predictive power independent of intention, such that in areas of high recycling, take-up might be higher even if other factors, such as convenience, and especially knowledge, are lacking.

Finally, there is a small negative direct effect from the acceptance of the norm to recycle, which appears to make little sense in the light of previous sections. However, if the overall effect on recycling behaviour is examined, it is seen that the acceptance of the norm to recycle is very small (0.009) when both direct and indirect paths are assessed. Hence, accepting that one will act if others do appears to be very much an aspiration rather than reality, and it appears that awareness of norms (perhaps more correctly termed acceptance of norms) has more impact.

Overall summary Recycling behaviour appears to be contingent on relatively few, but well fitted factors (beginning with the most significant first):

• Willingness to recycle (comprised of acceptance of norm to recycle, local waste knowledge, convenience/effort, active concern and house type).

And:

• Access to a recycling bin;
• Convenience/effort factors;

• Local waste knowledge;
• Awareness of norm.

This compact set of variables accounts for nearly four-fifths of the variance in recycling behaviour and therefore provides a good understanding of recycling behaviour in Exeter at the time of survey.

Minimisation, Reuse and Recycling Behaviour: Evaluation

The first point to be made here is that all three behaviours, despite two of them having the same intention scale, are fundamentally different in terms of the variables that predict them. Although minimisation and reuse behaviour are similar, differences are present, especially when the interaction between intention and behaviour is examined. Although the coefficient between willingness to act and behaviour is similar, there are more direct effects on behaviour for minimisation than for reuse. Indeed, these are all different. For those factors that influence intention *and* behaviour, these are also very different. For example, gender has an overall impact of 0.13 in the minimisation model, whereas this is merely an indirect effect of 0.02 in the reuse model. Conversely, experience has an overall effect of 0.08 in the reuse model (combining its direct and indirect effects), whereas in the minimisation model it only has an indirect effect of 0.02. Although these differences are small, they point to the fact that these behaviours are different.

One key area of difference, and this is especially so when recycling is compared, is that environmental values are more significant when minimisation is examined (overall effect of 'Human Priority' scale is 0.17, overall effect of Importance of Nature scale is 0.14). This reduces to an overall effect of 0.13 for the Importance of Nature scale and an indirect effect of only 0.04 for reuse. For recycling, the overall effect of the importance of nature scale is 0.05. Above it was suggested that such a pattern might reflect the importance of environmental values in shaping non-normative behaviour, but that this influence declines with socially 'normal' behaviour, like recycling.

Although the determinants of willingness to minimise or reuse obviously remain stable, there is a shift towards a different set of factors explaining reuse behaviour that emphasise convenience and effort, as well as response efficacy and intrinsic factors. The first of these is prominent in explaining recycling behaviour, and may suggest a shift from a values-knowledge-based prediction (as with minimisation) to factors that are more in common with socially normal behaviour. Clearly there is a shift from minimisation behaviour, relying on detailed knowledge, age cohort and a perception of a community spirit, to reuse behaviour, relying more on facilitating factors and efficacy factors.

This trend is completed when recycling intentions and behaviours are considered. Values play very little part, as do other forms of knowledge apart from specific recycling knowledge. Overall effects suggest that this knowledge, as well as enabling and motivating factors (having a bin and convenience/effort factors) combine with normative elements to provide a set of predictors distinct from those that explain reuse and minimisation behaviour. Above all, the data suggest that recycling is 'normal' (or 'normative'), and that willingness to undertake such behaviour and actually behaving in that way is a function of logistical factors rather than deeply held values or detailed knowledge (e.g. about the waste problem, its threat, policy, or environmental knowledge). Community factors and involvement in groups appears to be of no importance. Such a finding supports a thesis that would suggest that recycling is socially accepted as behaviour, but that minimisation, and to a lesser extent, reuse, is governed by more complex and deeper factors than merely specific knowledge or normative elements.

As a final note, it is interesting to see that household type, occupation and political affiliation did not have any impact. Indeed, few of the socio-demograhics were important, apart from age and gender. This provides an interesting comparison to the United States where the social bases of environmental concern (and behaviour) were thought to be crucial. Of course, this is one local study among many others, but it nevertheless provides food for thought and certainly legitimises the rationale for this study.

Conclusions

In terms of the salient statistical conclusions from this chapter there are a number of points that can be made. The first point to note is the fact that the conceptual framework used here provides an effective way in which to order certain environmental behaviours. By providing a conceptualisation, but making this neither an exploratory or confirmatory piece of work, the flexible nature of the technique used has been useful. It is clear that, as posited in the framework, willingness to do a certain action (here described as 'intention') has distinct antecedents from behaviour. There are also intervening variables that may act totally independently of intention or have both direct and indirect effects on behaviour. The strength of the 'intention-behaviour' relationship here also supports the use of this conceptual framework, and suggests that willingness to do certain activities is important, but not the only factor governing behaviour. Most crucially, however, the framework has permitted a large number of factors to be analysed in a flexible framework.

The second point to make is that there is considerable overlap in the results given in the brief bivariate sections and those alluded to above in the multivariate section. The clearest manifestation of this fact is that it was concluded in the bivariate section above that there were strong associations between intentions and behaviour. This has clearly been shown in the multivariate results. Multivariate analysis has provided a fascinating insight into the data set and has allowed clear patterning in the data. It is evident that minimisation, reuse and recycling behaviour are different in terms of their antecedents. Minimisation and reuse behaviour are more similar, but this tends to be with reference to intention rather than actual behaviour. Indeed, recycling behaviour has very unique antecedents. This finding has crucial and fundamental impacts for policy in the arena of waste management, which will be dealt with in Chapter Nine.

The major point to make from this entire analysis is that, taken as a whole, the quantitative data support heavily the notion that recycling behaviour is an accepted activity, which has certain links to environmental concern, but is predicted mostly by access to a kerbside recycling facility. Recycling can therefore be summarised as being affected by access to a recycling bin, awareness and acceptance of the norm to recycle (the latter acting via intention), feelings of convenience and enough space to store reyclables, as well as context-specific knowledge of where and what to recycle. Other factors act to maximise these variables, such as access to a car and use of various knowledge sources. The fundamental point here, however, is that recycling is not a value-based activity, but rather context-specific, relying on a good recycling network and the spin off of good publicity and minimising of structural and psychological barriers, such as convenience and space to store recyclables.

Contrasted with this finding is minimisation behaviour. Univariate statistics in Chapter Six demonstrated that this variable was very distinct from recycling in particular. This has been confirmed by the multivariate analyses. Minimisation behaviour is essentially values-based, there being direct effects on it from both environmental values factors, as well as indirect effects via willingness to minimise. What makes minimisation separate from recycling behaviour, as well as reuse behaviour, is the prominence of gender and age both as predictors of many of the other regressors, as well as having direct effects themselves. Clearly, older people and females are more willing and do minimise their waste. Additionally, being part of the community (in an active sense) also appears crucial. Minimisation behaviour obviously has the same predictors as reuse via willingness to minimise, and this points again to a knowledge, values and 'citizenship' element. A fundamental point from the policy point of view is that those who tend to be less willing to minimise (and reuse) tend to be those without a recycling bin, reflecting on the fact that having a recycling bin might have a negative impact on minimisation behaviour. Evidence also points to the fact that minimisers are more concerned with waste issues and the related problems. Fundamentally, what makes minimisation unique, therefore, is the reliance on gender and age as direct and indirect predictors.

Of course reuse behaviour has many of these antecedents, but what makes it unique is the fact that convenience and effort have a significant direct effect, along with factors emphasising response efficacy and intrinsic motivation (MR Scale). Reduced reliance on environmental values, points to the fact that although reuse behaviour has similar antecedents to that of minimisation behaviour, some logistical factors are of relevance, as well as the fact that reusing appears to engender feelings of satisfaction and that acting to reuse will have a significant positive effect on the environment. Indeed, recycling experience has a direct effect on reuse behaviour, whereas it has no such impact on minimisation actions.

The fact nevertheless remains that minimisation and reuse behaviour are similar in their antecedents, whereas recycling behaviour is fundamentally different. This fact supports Oskamp *et al.*'s (1991) assertion that environmental behaviour is diverse. Nevertheless, such a conclusion has not been reached by many in this area of research, most notably those in government, who treat the issue of waste as one entity, whereas as can be seen here, it is far from that.

Table 7.6 shows a summary of results from the statistical analysis.

Table 7.6 **Summary of factors influencing the three factorially-defined waste behaviours**

Process	Behavioural dimension		
	Minimisation	Reuse	Recycling
Behavioural intention	• Environmental values • Active concern (stated concern and moral obligation) • No kerbside bin • Problem and threat perception • Recycling experience • Citizenship factors	• Environmental values • Active concern (stated concern and moral obligation) • No kerbside bin • Problem and threat perception • Recycling experience Citizenship factors	• Acceptance of norm • Active concern (stated concern and moral obligation) • Local waste knowledge • Convenience and effort • House type
Behaviour	• Behavioural intention • Age (older groups) • Knowledge sources • Feelings of community/local democracy • Environmental values • Gender (female) • Policy knowledge	• Behavioural intention • Motivation to respond (response efficacy and intrinsic motivation) • Convenience and effort • Community group membership • Importance of Nature scale	• Behavioural intention • Kerbside bin • Local waste knowledge • Convenience and effort • Acceptance of norm

Stated relationships are positive unless indicated otherwise

8 A Framework for Advancing Theory and Policy?

Introduction

This chapter examines the efficacy of the framework and approach that was advocated in Chapter Five. The detailed results presented in the previous chapter are examined in the first instance and comparisons to past research in this field are made. The effectiveness of this research and its relatedness to the work presented here is examined. The chapter then moves on to consider how the literature-based framework developed in Chapter Five can be amended in order to take full account of the results presented in Chapter Seven. The amended framework is assessed and examined in terms of the likely efficacy of such a framework. Finally, the chapter examines the policy context and uses the results from Chapter Seven in order to arrive at practical policy recommendations that can be derived from the statistical analyses.

Previous Theories and Current Research

The extent to which the research outlined in Chapters Two to Four has efficacy within the framework developed in Chapter Five is questionable since the factor analyses presented in Chapter Seven have shown the alternative variables that can be derived in particular studies. This section examines the efficacy of each perspective in turn, proceeding first to assess the effectiveness of environmental values and then moving to examine the situational and psychological factors.

Environmental Values

The scales used to assess environmental values were constructed on the basis of a principal components factor analysis with orthogonal varimax rotation. Two factors emerged: 'Importance of Nature' and 'Human Priority' when protecting nature (Table 7.3). The former conformed to an ecocentric view of the environment, whilst the latter represented a human rationale and/or consideration when examining environmental protection, representing a slightly more technocentric view (O'Riordan, 1985). Nevertheless, higher scores on both scales measured enhanced levels of concern for the state of the environment. This is an important finding, since many researchers have utilised environmental values scales individually and have not sought to assess their internal dimensions. As Albrecht *et al.* (1982) and Kuhn and Jackson (1989) found with the New Environmental Paradigm (NEP), more than one dimension was involved. Although Albrect *et al.* found three dimensions in the NEP, it is interesting to note that there was a division between 'limits to growth' and 'man over nature' orientations. Clearly the ecocentric-technocentric continuum envisaged by O'Riordan (1985) still has some significance. Interestingly also, no dimension measuring apathy appeared to emerge from the factor analysis, even though this was a key finding of Thompson and Barton (1994). Nevertheless, as was reported by a number of authors (e.g. Dunlap and Van Lierre, 1978; Vining and Ebreo, 1990; Scott and Willits, 1994; Widegren, 1998) scores were generally high and positively skewed on values scales. As Figure 6.4 demonstrates, mean scores were high, and this was reflected in the breakdown in score distributions, with a large majority scoring 4 or 5 in agreement. Nevertheless, as Scott and Willits (1994) note, this is not necessarily reflected in the value-behaviour relationship.

Tables 7.1 demonstrates the bivariate relationship between environmental values as measured in this study and behaviour. The trend toward values having more impact on minimisation and reuse is significant. Nevertheless, although there is a definite trend, none of the correlation coefficients are high. Hence, Weigel and Weigel's (1978) finding of a high correlation between values and behaviour ($r = 0.62$) is not supported here. Nonetheless, as Tarrant and Cordell (1997) have stated, there is a problem with merely correlating values to behaviour, since socio-demographic (and other) variables might have a moderating effect on the relationship, ultimately making the apparent correlation either spurious or more significant than is actually the case. Hence, as with authors such as Oskamp *et al.* (1991) and Mainieri *et al.* (1997), the data were put into a regression analysis in order to asses the relative predictive validity of environmental values upon behaviour. Table 7.5 shows the results of these analyses. As can be seen, the direct effects on all the behaviours are reduced by this procedure, although as the bivariate analysis demonstrated, there was a greater direct effect upon minimisation and reuse than recycling behaviour.

Hence, although there is a weak effect, it appears that environmental values have only weak predictive power for minimisation behaviour and virtually none for recycling. This is consistent with the findings of Mainieri *et al.* (1997). The findings of Arbutnot (1977) who found no relationship cannot be supported, neither can those of Oskamp *et al.* (1991) who found a negative predictive power of environmental values, nor Steel (1996) who stated a large predictive power for the NEP on environmental action. Clearly, there is considerable disparity between different studies that use alternative measures of environmental values and behaviour. Hence, it appears that concluding whether environmental values have an impact upon environmental behaviour depends on the scales used, the degree of specificity and the analytical model used to assess the relationship. Certainly assessing empirical dimensions of scales and assessing these against other predictor variables is likely to filter some problems out. Nevertheless, it is likely to be the case that very strong correlation coefficients between variables are spurious and that there is indeed a value-action gap. Elucidating the nature of the 'gap' is the current problem. In the Exeter study it was the case that the gap depends on the specific behaviour being examined. Hence, although the logical process in Figure 5.2 is maintained, this is dependent on the behaviour being examined. Indeed, until a reliable measurement instrument for such environmental values can be developed, studies are likely to continue to report diverse results.

Situational Variables and Environmental Behaviour

Context: Kerbside Recycling Collections and Static Recycling Provision

The Exeter study operationalised context in two ways. Consistent with previous research (e.g. Berger, 1997; Derksen and Gartell, 1993; Guagnano, *et al.*, 1995) access to kerbside recycling was taken as a proxy for a very unique element of social context. Distance to the nearest recycling facility was used to assess the overall spatial provision available to each household.

Table 7.5 reveals that regression analysis showed no direct effect on behaviour and only a very poor fit to recycling intention. This is at odds with the univariate and bivariate data given by Ball and Lawson (1990) and Barr (1998). There was a clear relationship in these studies between provision and behaviour. However, in the case of Barr (1998), kerbside collections and non-kerbside collections were integrated to form scales.

This may explain the current weak correlations, since the current study sought to separate the effects of static and commingled kerbside recycling. This is seen when kerbside recycling and its effects are directly examined. Table 7.5 shows clearly that there are statistically significant differences between those who have access to commingled kerbside recycling and those who only have access to static provision. This has been universally found by all researchers, and would be surprising if not! Barr (1998) found that those with access to green box recycling at the kerbside in Oxfordshire had higher recycling scores overall. Derksen and Gartell (1993) demonstrated that Edmonton residents with access to a kerbside recycling bin were much more likely to recycle than those without such a bin. Guagnano *et al.* (1995) reported a similar result. Indeed, Berger (1997) demonstrated how adding social context as a predictor can override traditional hypotheses regarding environmental behaviour.

Clearly, the finding of the current research is very much in step with other work. Access to a kerbside recycling bin is very important in motivating recycling (although note not recycling intentions). Nevertheless, the anticipated spatial relationship did not appear and this is somewhat confusing, since it might have been anticipated that there would have been a relationship between distance travelled to a recycling site and behaviour. This may, however, be more to do with the arbitrary measurement instrument rather than the actual relationship.

Socio-demographics and Environmental Behaviour

Age   Table 7.5 shows that although age is a weak predictor of recycling, it is a good predictor of minimisation behaviour. Essentially, those in older age groups tend to minimise their waste more frequently. Given the paucity of the data concerning waste minimisation this cannot be placed in context, but does demonstrate the value of using regression techniques to assess the value of a variable.

This finding certainly opposes the results given by Weigel (1977), Sia *et al.* (1985) and Oskamp *et al.* (1991). It does support the findings of Baldassare and Katz (1992) and Derksen and Gartell (1993) who found that age had a significant, but weak predictive value for recycling and other environmental behaviours. In general, those who have assessed age in comparison to other variables have found that its effect has been less profound. Nevertheless, Oskamp *et al.* (1991) did find a negative relationship when this was undertaken and hence a cautionary note should be struck. On the other hand, Oskamp *et al.* did not report the effect of age with reference to anything but recycling behaviour. The finding that age is important in intervening between minimisation intention and behaviour is important and implies that older people are more likely to minimise as a course of habit. This is important to note in the development of theories of environmental behaviour.

Gender   Table 7.5 demonstrates that when gender is seen in the context of all other factors, women only tend to have higher scores than men for minimisation. This is another important finding, since Baldassare and Katz (1992) found that although gender predicted four environmental behaviours, it did not predict recycling at home. Indeed, Schahn and Holzer (1990) and Witherspoon and Martin (1992) both reported gender as important in modifying environmental behaviour across their behaviours. That conclusion cannot be drawn here, and because these authors did not categorise their behaviours there is no way of knowing what effect gender had on the empirical dimensions of their behaviour scales. Nonetheless, it appears that gender may have some explanatory power at some levels of behaviour (Baldassare and Katz, 1992).

This finding does, however, oppose Blocker and Eckberg's (1997) assertion that gender has no effect on environmentalism. It appears that focusing on the dimensions of environmental behaviour, as stated above, is likely to show that there are differences in the predictive value of gender.

Car access   As Table 7.5 shows, there is no effect of car access onto any of the behaviours. This may, however, be moderated by other factors and/or scales which may be affected by access to a motor vehicle.

House and household type   Table 7.5 shows the weak positive association between larger house type and recycling intention and behaviour. It therefore appears, as hypothesised here and by workers such as Berger (1997) and Derksen and Gartell (1993) that size of dwelling is important, although here it was only so with regard to recycling intentions. Indeed, its predictive value is less than other knowledge and convenience factors, implying that its effect may only be important when these influences are reduced.

In terms of household type, Table 7.5 shows that there is no significant effect on intentions or behaviour depending on the type of household that an individual is part of. This is at odds with workers such as Oskamp et al., (1991), Lasana (1992), Beger (1997) and Daneshvary et al. (1998) who all found differential relationships according to different family types.

Occupational Status, income, education and political affiliation Table 7.5 shows that there are no statistically significant differences in intention or behaviour scores according to respondent occupation, income, education or political affiliation. These findings are of importance since they demonstrate the lack of efficacy in certain socio-demographic factors that have previously been of significance in other studies.

Conclusion This study has shown that many of the previously cited factors influencing environmental behaviour have not been found important. Only age has a slight positive impact upon recycling behaviour. However, age and gender are both important in predicting minimisation behaviour. This is an important finding since it shows that examining different types of behaviour can bring quite different results regarding significant predictors. Nevertheless, the overall impact of the demographics of the sample was weak in all cases.

Knowledge

Chapter Three demonstrated the importance of knowledge in the prediction of environmental behaviour. Table 7.5 shows that local waste knowledge had a very significant direct effect on recycling behaviour and was the third most important predictor (after waste intention and having a kerbside recycling bin). However, given its direct effect on recycling intention, its overall effect on behaviour was increased (Figure 7.3).

   This is certainly a well-anticipated result and again shows the value of the methodology used. Nevertheless, a cautionary note must be struck here, since as Sia *et al.* (1985) have found, specific behavioural knowledge may only be important for predicting some environmental behaviours. Their study of environmental behaviour in Illinois demonstrated that on their index of behaviour, specific knowledge was placed below psychological factors when predicting action. Since the current research did not develop a specific knowledge measure for minimisation and reuse it cannot be said that specific knowledge does not have an impact upon these variables, and this consequently needs further development.

   Nevertheless, by following the recommendations of Schahn and Holzer (1990) a solution to this problem may have presented itself. By measuring 'Abstract' knowledge, the current research developed three measures of environmental knowledge, based around general environmental awareness, specific waste awareness and policy awareness. Table 7.5 and Figures 7.1 and 7.2 show the effect of these factors. Policy knowledge plays an important intervening role between intention and behaviour for minimisation. Environmental knowledge has a significant, if minor, impact upon minimisation and reuse intentions. Hence, unlike recycling, it appears that minimisation and reuse behaviour is affected by more abstract knowledge. This is an important finding since it implies that context specific knowledge is not always the best predictor of behaviour (Hines *et al.*, 1987). However, the effects overall are small and it should be noted that knowledge of waste issues had not effect at all.

Finally, with regard to knowledge sources, Table 7.5 shows that although knowledge sources have a significant impact upon recycling intentions, this is minor when compared to other factors. However,, they have a very significant direct effect on minimisation behaviour. Thus it appears that enhanced use of diverse knowledge sources has an important effect on reported minimisation behaviour. This is an important finding since this variable has not been tested previously and implies that those who use a variety of knowledge sources to gain information on waste issues are those who tend to minimise their waste more.

Experience

The current study operationalised behavioural experience by asking respondents with a kerbside recycling bin whether they had recycled before receiving that facility. This therefore acted as a direct measure of experience for recycling as well as a more general behavioural factor with which to compare with minimisation and reuse behaviour. Table 7.5 reveals that 'experience' only has a moderate overall effect upon reuse behaviour and a small effect on minimisation/reuse intentions. There is no effect on recycling behaviour. This is probably because recycling is well defined by other principal variables such as specific knowledge and kerbside collection schemes. Some behavioural experience therefore seems important in shaping reuse and minimisation behaviour. This is in line more with the work of Oskamp *et al.*, (1991) who found differential effects on different forms of recycling behaviour according to the specificity of these actions and the experience gained. Daneshvary *et al.*, (1998) also supported the notion that wider experience reinforces certain behaviours. However, the current research does not lend evidence to the idea that recycling behaviour will lead to further action (as with Luyben and Bailey, 1979, and Goldenhar and Connell, 1992-1993). This is due to the fact that other factors in the regression analysis proved more predictive and implies that experience may be more useful as a broader concept than related to a specific set of actions.

Situational Variables: Conclusion

The current study has demonstrated a number of key differences between the literature and the study findings. First, the prominence assigned to socio-demographics is unfounded and although age and gender have an intervening effect between intention and minimisation behaviour, this is an exceptional case. As Berger (1997) and Tarrant and Cordell (1997) have stated, demographic effects can be misleading, especially when examining bivariate relationships, since they may in fact form spurious rather than true relationships with behaviour. Second, the study has found that by distinguishing between empirical dimensions of behaviour, the effects of certain variables are altered. This is especially noticeable with regard to knowledge and waste behaviour. Although (as expected) the effect of contextual knowledge was important upon recycling behaviour, 'Abstract' awareness of environmental issues and policy initiatives both had important effects. This is in contrast to Schahn and Holzer's (1990) view that only specific knowledge would be important, and supports the view of Hines *et al.*, (1987) that general knowledge can be important in predicting environmental behaviour. Indeed, knowledge sources had differential direct and indirect effects on different types of behaviour. Third, it appears that the role of behavioural experience is important at a general level, rather than linked to specific actions. This supports work by Daneshvary *et al.* (1998) who found that kerbside recycling of textiles could be predicted by other kerbside recycling. However, it contradicts the work of Luyben and Bailey (1979) who contended that direct behavioural experience shapes future behaviour. Hence, by using aggregate and multiple regression techniques the current research has enabled a full test of the hypotheses concerning environmental behaviour and has enabled the data to be sorted accordingly.

Psychological Factors and Environmental Behaviour

This section outlines the combined effects of the psychological variables outlined in Figure 5.2. Schwartz's (1977) model of normative influences on altruism is assessed first, and the series of variables alluded to in Chapter Four are then discussed in the context of previous work.

The Schwartz Model and Waste Behaviour

Figure 4.1 outlines the dynamics of Schwartz's (1977) seminal work. In common with other studies in this field, the components used to evaluate the model were factor analysed in order to elucidate their empirical dimensions. Along with the other psychological variables, these are given in Table 7.4. This is the first step to assess the validity of the Schwartz model, since other studies of the model have sought to test the model on its own. This research seeks to assess the model as compared to a broader framework of behaviour.

As Chapter Four demonstrated, previous authors have attempted to streamline the Schwartz model for their own needs and in a similar vein, this research measured the principal variables within the framework. Essentially, 'Awareness of need', 'Perception that relevant actions are available' (including recognition that one can provide relief), 'Ascription of responsibility', 'Personal norms', 'Assessment of personal costs' and 'Behaviour' were measured. Table 7.4 demonstrates that with regard to the first five of these factors, there were mixed results. 'Awareness of need' is represented by Factor 5 (Problem/threat). This incorporates notions of a problem and a threat to the self. A need is therefore identified here. Factor 3 (Beliefs) represents 'Perception that there are actions that can provide relief' (belief outcome of the behaviour). Factor 9 (Citizenship) represents the 'Ascription of responsibility', although as can be seen in the Table, the second item here loads on Factor 2 (Active concern). Factor 2 itself partly represents the 'Personal norm' to recycle, and incorporates notions of moral obligation to act and concern for the object in need. Factor 1 (Convenience/effort) represents the 'Assessment of consequences', although response efficacy loaded onto Factor 4 (Motivation to respond). Finally, as already mentioned, Table 7.2 shows that there were three empirical dimensions to the behaviour items.

As can be seen from Table 7.4 this initial exercise was not entirely successful, since some factors have other variables included in them that were not specifically relevant to the Schwartz model. However, this shows the value of the technique used here and the problems of merely collecting data to prove or disprove a particular theory. This is an important finding since it implies that previously well-established factors are less defined here and may constitute a wider range of influences. However, given that factors could be roughly identified that link to the model, how does the causal structure of the Schwartz framework withstand the multivariate analysis used here?

The limitations of space and the need for brevity do not allow a detailed analysis of the workings of the Schwartz model in this case. This is undertaken in Barr (2001) and can be summarised as follows. Essentially, using the constructs as defined above, the model only functions partially for reuse behaviour. The major difficulty with linking the Schwartz variables empirically is that they have loaded differently in the factor analysis undertaken in this study, probably due to the large number of factors involved. There is evidence, as mentioned above, that reuse might be characterised as an 'altruistic' behaviour, as the process for the Schwartz model given in Figure 4.1 functions well. This assertion is supported below to some extent. However, overall, given the alternative 'loading' of factors to that intended by authors such as Hopper and Nielsen (1991), it cannot be said that waste management actions are altruistic.

Satisfactions and Intrinsic Motives to Manage Waste

In order to test the thesis of De Young (1986) that intrinsic motives to manage waste carefully, such as satisfaction from behaving accordingly, had an effect upon actual action, two items were placed within the questionnaire that assessed individual satisfaction on the one hand and attitudes to extrinsic motivations on the other. As can be seen from Table 7.4 these statements loaded onto the same factor as those with response efficacy. However, if this is accepted as one factor incorporating this variable, then Table 7.5 shows clearly that satisfaction with environmental behaviour, rejection of extrinsic motives, along with a belief that one's actions will be significant, has a large direct effect on reuse behaviour. This indicates that as Werner and Makela (1998) found, people who enjoy reusing activities are more likely to undertake them. This does of course only work for reuse behaviour and implies that satisfaction plays a minor part in forming minimisation and recycling behaviours. This is in contrast to De Young's (1986) work, as well as that of Oskamp *et al.* (1991) and Gamba and Oskamp (1994).

Of course, the nature of the factor analysis demonstrates that the intrinsic motivation and response efficacy factors are empirically similar and this would imply that some interaction between these two constructs is apparent, perhaps manifested in the idea that satisfaction is derived from a belief that one's action will achieve a certain goal, or the knowledge that actions achieve certain goals brings forth satisfaction.

Subjective Norms and Waste Behaviour

Figure 7.3 clearly shows the vital influence of both an acceptance of a given norm to act and the awareness overall of this norm. This is consistent with Gamba and Oskamp's (1994) finding that social 'pressure' was important, but not crucial, in shaping recycling behaviour. However, Oskamp *et al.* (1991) and Tucker (1999b) cannot be supported in their assertions that normative influences were vital. Indeed, Vining and Ebreo's (1990) assertion of no difference in normative influences is also not supported here. As Figure 7.3 shows, awareness of others' recycling is likely to override a given intention to recycle. In other words, even if it was stated that there was not a willingness to recycle on behalf of an individual, it might still be undertaken if others around were undertaking this action. This very awareness, however, is likely to base on certain situational factors.

Environmental Threat and Waste Behaviour

In Chapter Four it was noted that Baldassare and Katz (1992) found that environmental threat predicted their four environmental behaviours far better than the cohort of socio-demographic factors often used to predict such actions. Table 7.4 shows 'Threat' was empirically related to notions of the waste problem. It is logical that these two factors would be linked, since admission that waste was a problem would imply some sort of personalisation of that problem, by reference to a 'Threat'. Table 7.5 shows that this factor was important in shaping minimisation and reuse intentions, and as Figures 7.1 and 7.2 show, had a small indirect effect upon those behaviours. This result does not support the assertion of Baldassare and Katz (1992) or Segun *et al.* (1998) that threat is a direct predictor of environmental behaviour. Certainly the former authors are right to state that threat has an effect, but this is confined to a willingness to act, not action. Indeed, it is confined to minimisation and reuse behaviour.

Efficacy and Waste Behaviour

Table 7.4 shows that these constructs (self and response efficacy) expectedly load onto different factors. Self efficacy loads with logistical barriers, whilst response efficacy loads with intrinsic motivation. Self efficacy, along with logistical factors, predicts both recycling and reuse behaviour and orientation to recycle. This is in line with the findings of

Chan (1998) in Hong Kong. It also lends support to the *Theory of Planned Behaviour* (Ajzen, 1991; Ajzen and Madden, 1986; Kantola *et al.*, 1992). With regard to response efficacy, as already noted above, the main direct effect along with intrinsic motives to act is upon reuse behaviour, with additional indirect effects via the 'Active concern' and 'Beliefs' scales. This again would imply that believing in one's efficacy to solve environmental problems bolsters concern and beliefs that waste management practices are effective, which in turn enhance reuse behaviour. This supports Oskamp *et al.* (1991) in their finding that response efficacy only had an indirect impact upon recycling behaviour, and does not support the notion of Arbutnot (1977) or Becker *et al.* (1981) that response efficacy has a direct effect upon behaviour.

Logistics and Waste Behaviour

Table 7.5 shows that the primary predictive power of this factor was in the prediction of recycling and reuse behaviour, as well as recycling intention. Studies that have incorporated such variables have found that such factors are indeed empirically linked (e.g. Gamba and Oskamp, 1994; Lasana, 1992, 1993; Howentsine, 1993). Since studies have been limited there is no evidence from this past research that there are any effects upon behaviour apart from recycling. However, all the studies cited demonstrate that convenience and effort factors are usually significant in predicting recycling behaviour. The current research places a high prominence to such factors, and it is the contention here that unlike the work of Gamba and Oskamp (1994) such factors are among the primary determinants of both recycling intentions and behaviour (Table 7.5, Figure 7.3. This is not the case with reuse behaviour, where convenience factors intervene to moderate the relationship between intention and behaviour. It is highly likely that those who are willing to reuse waste and have room to do so will have enhanced reuse scores as shown in Figure 7.2. This is another important finding from the current work.

Environmental Citizenship and Waste Behaviour

As shown in Chapter Four, little use of citizenship variables has been undertaken in attitude research, and therefore this is somewhat of an exploratory section. However, it was hypothesised that those who were more likely to act in environmentally positive ways were those who believed they had environmental rights on the one hand, and environmental responsibilities on the other. Indeed, they would be more involved in the community and believe in the efficacy of local democracy.

Table 7.4 shows that the community involvement and local democracy items have loaded onto one factor, with the rights and one responsibility onto another factor. Table 7.5 shows that the effect of these new factors is more significant. The 'Community/democracy' factor has a direct effect on minimisation behaviour, indicating wider community participation and feelings of local democracy enhance minimisation behaviour beyond one's intention, implying that a sense of civicness enhances a stronger willingness to take part. However, rights and responsibilities act on both minimisation and reuse orientation such that the 'Citizenship' variable acts well to change willingness to act, but not action.

This is an important finding. It would imply that to a certain extent, people who believe in strong environmental rights and responsibilities are more willing to minimise and reuse their waste. However, the disparity between this willingness and action is explained by different factors according to the behaviour being examined. In the case of minimisation, belief in local democracy and community involvement improves the disparity. What is of interest also is the fact that, as was noted in Chapter Seven, since both minimisation and reuse behaviours are undertaken on a less frequent and different basis than recycling, it appears that the traits of environmental citizenship, such as rights and responsibilities, are those that in part support reuse and minimisation orientation. This being the case it supports the notion of writers on this subject (e.g. Selman, 1996) that such values are held by relatively few. The more established behaviours, such as recycling, are not values-based and rely on contextual and basic psychological factors. The more marginal behaviours rely more upon values and concern-based traits, such as notions of citizenship. This is a further important finding of the current research.

Summary

The most salient point to draw from the results is that there are both dimensional and correlational differences between the types of psychological variables on the one hand and their effect on behaviour on the other. This has implications for both theories pertaining to likely influences on environmental behaviour and the effect of these on behaviour. For example, demonstrating that within a complex study of waste behaviour that the Schwartz model does not necessarily function as expected has significant implications for users of this framework, since it would imply that although the variables within the model are logical antecedents of action, their position in the process is questionable.

Evaluation of Factors that Influence Waste Behaviour in the Context of the Existing Literature

Within the context of the literature there are a number of fundamental points that must be made concerning the efficacy of the factors investigated in the current research. As an overall point, which will be elaborated below, the analytical and organisational framework of any piece of research such as this, inevitably has impacts upon the efficacy of certain factors within the strategy. However, by using a flexible and carefully analysed framework, the detailed comparisons summarised below ensure that the plethora of variables within the literature have been logically examined. In essence:

- The importance of analysing environmental values in terms of their empirical dimensions as well as their effect upon environmental behaviour was found to be crucial. Environmental values, both in ecocentric and more technocentric terms, were important in the prediction of minimisation and reuse behaviour, but not recycling behaviour. This is in contrast to those who have found a relationship between environmental values and recycling activity;

- Examining context with regard to kerbside recycling was very useful and as with the few other studies that have examined this variable, access to kerbside recycling was found to be crucial when predicting recycling behaviour. This variable is clearly under-investigated and may well prove certain previously cited relationships spurious;

- Examining context with regard to spatial variables met with little success. Unlike previous research, no relationship was found between static recycling site provision and behaviour. The measurement instrument may be to blame, although perceived distance, as measured in this and other studies may be a better proxy;

- Very few socio-demographic variables correlated to behaviour in any sense. This is in contrast to the literature and it is thought that this is due to the sorting of variables that this technique allows. Previous studies may have been registering spurious relationships. Nevertheless, age and gender were important in enhancing minimisation behaviour.

This has commonly been found with reference to recycling, but it is important to note its prominence with minimisation, as opposed to recycling;

- Experience is an important indirect predictor of reuse and minimisation behaviour. This implies that research that stresses the relevance of general behavioural indicators is more likely to see the impact of experience than using direct experience of that behaviour as a predictor;

- The Schwartz (1977) model of altruistic motivation was only partially successful. The variables used to assess the model, some from previous studies, did not load onto all the factors that were expected. The causal structure only partly worked for reuse behaviour. It is argued here that whilst the elements of the model have value (e.g. awareness of problems, ascription of responsibility) the structure itself does not take into account the many other variables that affect behaviour;

- Intrinsic motives to recycle (satisfaction) were only directly related to reuse behaviour. This implies a role, but less significant than in previous studies, for intrinsic motivation. These were empirically related to response efficacy, implying that satisfaction and beliefs in efficacious outcomes are similar;

- Subjective norms (e.g. awareness of others recycling) acted to enhance both recycling intentions and behaviour. They had little effect on minimisation and reuse behaviour. The direct effect on behaviour was indicative of some studies, although not in line with Fishbein and Ajzen's (1975) TRA;

- Environmental threat had significant indirect effects on minimisation and reuse intentions, but no effect on recycling. This is in contrast to studies that have assessed this variable and found it had direct effects. This was empirically linked to perception of the waste problem, indicating that people may perceive the waste problem as one that is easily related to the self;

- Self-efficacy (beliefs in personal ability to act) had direct effects upon recycling behaviour and intention, as with other research in this area. This relationship was linked empirically to logistical factors such as convenience and ease of acting. This implies personal efficacy is linked to basic logistical factors governing behaviour;

- Response efficacy (belief in personal actions to have a valuable effect) had direct effects on reuse behaviour, but in contrast to other research, not on recycling behaviour. This was empirically linked to intrinsic motives to recycle (see above);

• Logistical factors (e.g. time to recycle) were important in predicting recycling behaviour. This is very consistent with other research. They also had a direct effect on reuse, which is an important finding. These were empirically linked to self efficacy (see above);

• Environmental citizenship variables only had effects on minimisation and reuse behaviour. This would indicate that those who expressed elements of 'environmental citizenship' (rights *and* responsibilities) were more likely to undertake activities higher up the waste hierarchy. Such a conclusion would support the notion that citizenship values are important in influencing behaviours hat are also influenced by other sets of fundamental values (e.g. environmental values; see above).

Table 8.1 below shows the salient findings of the current research in the context of the literature reviewed in Chapters Two to Four and the current chapter.

Therefore, some of the existing literature's findings are supported. However, as has been noted in Chapter Seven, the three behaviours have different antecedents. In the case of this review it has been found that some of the determinants of recycling have been discounted, whereas some of the previously found determinants of recycling have been found as important predictors of reuse and minimisation behaviour. Inevitably each study area has its own unique circumstances that will affect the results.

However, overall it must be said that this approach has enabled the empirical dimensions of the principal determinants of waste behaviour to be examined in a way that has sought to test those theories and ideas put forward in the literature. Indeed, in examining these empirical determinants it has been shown that many of the variables linked to environmental behaviour are measuring different constructs than were envisaged. This is important because it may be possible that certain factors, for example those in the Schwartz model, are not always as well defined in larger studies. Hence, although the comparisons made above use the variables measured for specific notions within the literature, it has been shown that there is a need to rethink these factors.

Table 8.1   Summary findings of the current research and relevance to existing work

Variable	Study findings	Comparison to previous research
Environmental values	• Two dimensions: ecocentric and human orientated • Important in influencing minimisation and reuse behaviour, but not recycling	• Previous work supports the use of multi-factor scales • Previous research suggests important effects on recycling behaviour
Situational variables	• Context was important in shaping recycling behaviour • Spatial differentiation had little effect • Few socio-demographic variables influenced behaviour • Importance of behaviour specific (to recycling) and general environmental and policy knowledge (to minimisation/reuse) • Experience of importance to environmental behaviour in general	• Importance of kerbside collections in previous work • Previous research suggests significant spatial effects • Previous work showed more impact of demographics • Supports both those who assert behaviour specific and general knowledge is important. Need to examine knowledge in context • Supports notion that experience can have significant effects
Psychological variables	• The Schwartz model of altruism contained partially different empirical dimensions working only for reuse • Intrinsic motivation, linked empirically to response efficacy, was only linked directly to reuse behaviour • Subjective norms important in predicting recycling behaviour • Environmental threat, linked empirically to problem perception, had indirect effects on reuse and minimisation behaviour, but not recycling • Self efficacy was empirically linked to logistical factors • Response efficacy, linked empirically with intrinsic motivation, had direct links to reuse behaviour, but not minimisation and recycling • Logistical factors, empirically linked to self efficacy, had a significant effect on recycling • Environmental citizenship had important effects on minimisation and reuse behaviour, but not recycling	• Contrasts to previous research where empirical dimensions were confirmed for recycling • Previous research has found intrinsic motivation more important in predicting recycling behaviour and empirically unique • Supports research on normative recycling action • Previous research suggests this link should be direct and to recycling and separated from problem perception • See 'logistics' below • Previous research has found stronger links between outcome beliefs and recycling, and has treated it as a unique variable • Supports previous research on logistical factors and recycling • Makes an important contribution to previous qualitative research and 'rights and responsibilities'

The Conceptual Framework: An Addition to Understanding

Chapters Two to Four outlined the vast literature covering this area of study. However, it is now necessary to briefly evaluate what contribution this type of framework can add to our understanding of human behaviour in the environmental realm. Initially, the TRA is re-evaluated in the context of the current research and the crucial role of behavioural intention as a predictor of action is assessed. The overall efficacy of the framework is then discussed in the context of the additions to understanding that it provides.

The TRA/TPB: A Good Basis for the Framework?

The development of the conceptual framework used in this research is based upon the *Theory of Reasoned Action* (TRA) and *Theory of Planned Behaviour* (TPB) developed by Fishbein and Ajzen (1975), Ajzen and Madden (1986) and Ajzen (1991). It was argued through Chapter Two that whilst the TRA/TPB provided a useful skeletal framework, previous research both on the TRA/TPB and the welter of other work into environmental behaviour required an expansion of the framework to encapsulate the 'new' variables in a flexible but theoretically grounded version of the TRA/TPB. Figure 5.2 shows this conceptualisation and it is seen that the logical outcome of behaviour is predicted by behavioural intention, as in the TRA/TPB. However, unlike the TRA/TPB, the antecedents of behavioural intention are threefold. Using the evidence from both studies of the TRA/TPB and other research, it is argued that logically the theoretical precursors to behavioural intention are an individual's environmental values. However, behavioural intention is also predicted by two other fundamental 'extraneous' factors, on the one hand, situational variables, and on the other, psychological variables. As Ajzen and Madden (1986) acknowledged, behaviour itself is predicted well directly by perceived behavioural control, and the framework given here extends this premise to make it clear that both the situational and psychological variables intervene in the intention-behaviour relationship.

By using an extended TRA in such a way, there are a number of key points that need to be made. First, in line with Fishbein and Ajzen (1975), the framework provided evidence that there was indeed a strong intention-behaviour link. Chapter Seven gives the Spearman correlation coefficients for behavioural intention related to behaviour. Although there is variance here, the reuse and recycling items are particularly strong. Figures 7.1 to

7.3 show the strength of the relationship between intention and behaviour for all the empirically related behaviours, the strongest being for recycling. This demonstrates what previous workers have found. The intention-behaviour relationship given in the current study was better than that achieved by Ajzen and Madden (1986), Goldernhar and Connell (1992-1993), Boldero (1995), Taylor and Todd (1997) and Chan (1998). Jones (1990) and Kaiser *et al.* (1999) achieved better levels of correspondence. Nevertheless, it appears that as this study has shown, there is a fundamental relationship between intention and behaviour that is stronger than any other predictor of behaviour. However, as a second point, it is also apparent from the work reviewed and results in this study, that these forms of behaviour are not merely a reflection of that intention. Figures 7.1 to 7.3 show this with varying complexity. The various authors who have argued that there is this intervention are thus supported (Goldenahr and Connell, 1992-1993; Boldero, 1995; Schultz and Oskamp, 1996; Taylor and Todd, 1997; Kaiser *et al.*, 1999; Lam, 1999).

Finally, and as is apparent from the preceding chapters, the number of variables within the TRA, but more to the point, the way these are conceptualised, have led authors to add variables which have on the whole supported the overall conceptualisation given here. As Boldero (1995) has demonstrated with regard to newspaper recycling, the more flexible the model, and the more variables that can be accommodated, the more insight is gained into waste behaviour. Using stepwise regression she found that the original attributes in the Fshbein-Ajzen model were redundant as compared to certain structural and other psychological factors.

Therefore, the TRA/TPB concept is a useful skeleton on which to place what is a flexible and complex framework for understanding intentions and behaviour in the context of a large number of behaviours. Other work has demonstrated the value of the intention-behaviour relationship (of key importance to policy-makers, as well as theorists), the importance of recognising that other factors affect behaviour directly, and that these factors are diverse in nature.

Contribution to Understanding

The framework used here offers an original application in a UK setting and an updated understanding of the determinants of environmental behaviour. Through the process outlined above, including the qualitative analysis, an amended conceptual framework has been devised (Figure 8.1). This amended framework makes explicit the variables found within this study to be important in influencing intention and behaviour. It is argued that this offers a valuable contribution to the debate concerning environmental behaviour and a primer to organising research strategies. This can be summarised as:

• A logical theoretical framework of values-intentions-behaviour that places the fundamental 'intention-behaviour' (or 'value-action') relationship at its heart;

• A recognition of a large number of 'new' variables that impact on intention and the intention-behaviour relationship. These have been carefully categorised and grouped; and

• The possibility that such a framework could be used in other areas of environmental behaviour research.

It is argued that the part-confirmatory and part-explanatory approach offers future researchers further scope for wider research, as well as addressing Bagozzi's (1992) concern that such research would lose the theoretical grounding of attitude studies. It also offers the opportunity for others interested in this field of study, not merely social psychologists, to use their expertise in this field.

The research strategy has also shown that previous studies that have sought to relate specific variables to behaviour may have been measuring more complex constructs than previously thought. Previous sections have demonstrated how although some factors comprised items that were expected, other were mixed. This was of particular relevance to the Schwarz model. Table 7.4 shows that some factors were easy to identify (Beliefs, Subjective Norms, Community elements, Citizenship), yet others had mixed components, such as the 'Convenience/Effort' scale, which included the self-efficacy items. This obviously had more specific relevance to logistical considerations as opposed to feelings of personal inability to undertake the behaviour. Similarly, the 'Active concern' scale had statements referring to moral obligation, concern, time and responsibility. Clearly, these concepts are linked in some way and may express a more general, social concern than alluded to in previous research. The 'Problem/threat' scale is simpler to assess and might indicate that researchers need to consider that the waste problem is more closely held and related to the self than simply an abstract assessment of the situation.

Figure 8.1 Amended conceptual framework

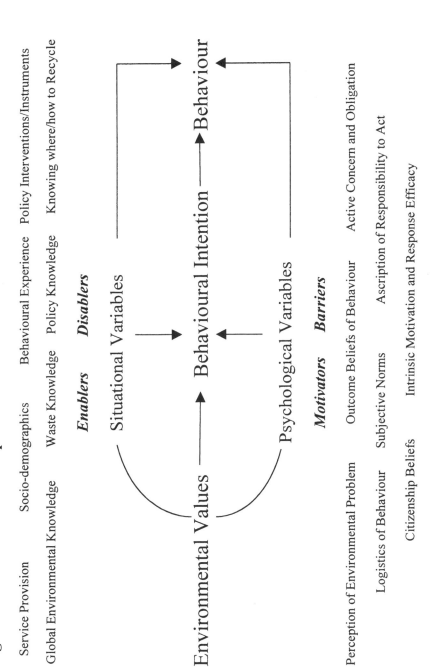

Policy Recommendations

This section examines the practical implications of the results described above in for the city of Exeter. Table 8.2 at the end of this section contains a resume of the recommendations. These practical policy changes and amendments are drawn from the multivariate section in Chapter Seven. Recycling behaviour is examined first, since this is thought to be the simplest behaviour about which to make recommendations. Minimisation and reuse behaviour are then dealt with and a range of policies are outlined that could, on the basis of this research, make a difference, and ensure that waste management by individuals moves up, rather than down, the waste hierarchy.

Optimising Recycling Behaviour

Reference to Figure 7.3 demonstrates the framework of recycling behaviour found for residents in Exeter. It is immediately apparent that policy recommendations should deal with two conceptually different areas when encouraging recycling behaviour. Initially, there is a need to change people's underlying willingness to recycle, since this plays a major role in shaping actual behaviour. Only after this can behaviour itself be examined.

Willingness to recycle A large *situational* influence on willingness to recycle is 'Local waste knowledge'. This of course has effects on both intention and behaviour, but will be dealt with here. Figure 7.3 shows that this uncertainty is likely to reduce both a willingness to recycle and also act as a barrier for those who want to recycle. Clearly, the Local Authority needs to move beyond issuing information on where to recycle, which it does already, and deal with how each material must be presented for recycling. More technical information such as this could be posted on official letters and documentation, such as Council Tax bills. Space could also be used in local free newspapers which are delivered to every home in the city. Because this would entail considerable more information than snappy messages about the need to recycle, use of radio, TV and general print media is thought inappropriate for this publicity. Nevertheless, the information should be concise and above all clear.

It can be seen from Figure 7.3 that a major *psychological* determinant of intentions to recycling is the 'Active Concern' scale. This scale contained items from the questionnaire that emphasised a concern for waste issues and an obligation to manage waste carefully. It therefore follows that making people feel more willing to recycle will involve more awareness of the waste problem and more importantly, a message that it is everyone's responsibility. A key policy recommendation, therefore, is to enhance existing publicity campaigns to go beyond exhorting participation by telling residents that recycling is important, perhaps by using 'shock tactics' and outlining predictions of the likely outcome of not managing waste effectively. Of course, such techniques may have short lived effects, but may be stimulus for behaviours that might be maintained by other predictors of behaviour, such as intrinsic motivation. In terms of encouraging personal environmental responsibility, continued emphasis on the efficacy of each person's action, with examples of how this has been achieved, should begin to engender more responsible intentions. Whether this has been achieved in the *Are You Doing Your Bit?* campaign is uncertain, but these tactics are the same as being recommended here. In terms of dissemination, and this stands for all areas in this section when publicity is discussed, the use of local press, TV, radio and posters must be used in order to capture as wide an audience as possible and to give across a uniform message.

A further *psychological* variable that has an impact on both intentions to recycling and actual behaviour is the 'Convenience and Effort' scale (Figure 7.3). This scale contained items in the questionnaire that dealt with the convenience of recycling, the space required to store recyclables and the difficulty perceived in recycling. There are a number of recommendations that can be made here. First, a large predictor of this scale was access to a kerbside recycling bin, and therefore a key recommendation of this research would be to ensure city-wide coverage of the 'Recycle from Home' scheme as soon as practicable. Second, in recognition that this may not be forthcoming in the near future, other recommendations can be given for those without access to such a bin. However, perhaps the Council should look at enhancing convenience further by introducing uniformity across the static sites. For example, ensuring that all supermarkets take all recyclables, reviewing the current vehicle height restriction at recycling centres that excludes many with taller vehicles, and ensuring that all recycling sites have a minimum of, say, three recyclables that are uniform across the city (e.g. glass, newspaper and drinks cans). Many people are evidently confused about the diversity of the service offered. A uniform system would help greatly, and should not engage extra funds, but rather tightening up of the system. Third, and related to the issue of local waste knowledge discussed above, promotional literature needs to be very clear about where such sites are and exactly what can be recycled.

Finally here, it is seen that the 'Acceptance of the norm to recycle' has a large impact on recycling intentions. This is dealt with below in relation to the 'Awareness of norm to recycle', but suffice to state here that norms will be reinforced by the visibility of recycling as an activity and will be supported by the recommendations made above.

Recycling behaviour Having dealt with intention, it is of course necessary to look at behaviour. As mentioned above, 'Local waste knowledge' and 'Convenience and Effort' in Figure 7.3 have important direct effects on behaviour, and the recommendations above dealt with this. However, as a further recommendation for the extension of this scheme, it is recommended that the Local Authority examine the idea of infrequent glass collections (which cannot be placed in the green recycling bin) for those who don't find getting to a bottle bank easy. Indeed, resource to Figure 7.3 shows that those living in lager house types are more willing to recycle and therefore there may be scope to examine alternative 'kerbside' collections for those living in flats or terraced housing, such as communal street recycling.

Finally, there is the issue of the 'Awareness of norm to recycle' as shown in Figure 7.3. Clearly, people who are aware of those around them recycling feel more compelled to do it themselves. This should be operationalised by simply seeing the growing number of people recycling. However, one suggestion from Hopper and Nielsen (1991) is the use of community representatives, or 'Block Leaders' to enhance behaviour. Dissemination of information and the social example set by the leader has been shown to improve recycling in the United States. However, how this would work here is not clear. The Council could examine this prospect, but given the efficacy of the other factors in the recycling framework, this may be unnecessary.

A note should be made here about the suggestions put forward in *Less Waste: More Value* (DETR, 1998) that are not recommended. The use of variable charging schemes, fining for misuse and taxation to increase recycling rates are all regressive. They reduce the intrinsic motive to recycle (De Young, 1986) and above all they do not appear to be necessary. By imposing an extrinsic incentive to recycle (e.g. a lower tax bill) the behaviour is not necessarily internalised, but is somewhat begrudged. As shown in this research, people who are willing to recycle generally do. Therefore, emphasis needs to be placed upon awareness of waste issues and making recycling simpler to do. These regressive economic actions are strongly opposed by the findings of this research.

Encouraging Minimisation and Reuse

Since the predictors of intentions towards minimisation and reuse are the same, these are dealt with below as one set of intentions. Minimisation and reuse behaviour are then dealt with separately. The crucial point to note, before continuing, is that these two behaviours are less well explained by the frameworks used here, and hence any policy recommendations must be treated with caution. However, the data that were analysed suggested definite paths and this offers some hope for positive action. Reference to Figures 7.1 and 7.2 should be used in conjunction with the following text throughout this section.

Willingness to minimise and reuse As can be seen from Figures 7.1 and 7.2, *environmental values* are crucial when considering minimisation and reuse intentions. Changing environmental values is not simple! However, since those measured here emphasise the importance of nature and the human reliance on nature when considering future development, it is thought vital that existing campaigns, such as Local Agenda 21, be strengthened to bring home the general message of environmental sustainability issues. Again, these must be kept separate from campaigns looking at the problem of waste. Messages must not be confused. Nevertheless, value change is very gradual, but significant progress on internalisation of sustainability messages will eventually play its part in changing intentions towards waste minimisation and reuse. Given this prediction, awareness campaigns about the importance of minimising and reusing must be the priority, as well as focusing on those with kerbside recycling bins.

Two major *situational* predictors of minimisation and reuse intentions are 'Experience of recycling before having a kerbside bin' and 'Access to a kerbside recycling bin'. This apparent statistical anomaly was dealt with in Chapter Seven. It was suggested there that those with a kerbside recycling bin were less willing to minimise and reuse since they already 'did their bit'. However, of those with a kerbside recycling bin, those who had experience of recycling were more likely to recycle. Since this proportion was small, it was concluded that the overall effect of access to a kerbside recycling bin was negative. Hopefully, therefore, as the 'Recycle from Home' scheme expands, those with recycling experience will minimise and reuse more. However, for those for whom this is not the case, and those already with a kerbside recycling bin, there are crucial considerations that have to be made about how to change intentions to reducing and reusing waste. It is clear that a focused campaign, of a similar vein to the general awareness one outlined above, needs to be undertaken. This should focus on those on the 'Recycle from Home' scheme, as well as those about to enrol on it. This should focus in particular on the importance of reducing and reusing waste, perhaps with the overall benefit of reducing the rubbish being kept in the bins. Emphasis on the fact that recycling is a last resort for rubbish is crucial, and that reducing waste is easy. Because this campaign would be more focused, more detailed information could be sent concerning how to minimise and reuse. The messages promoting these two must, as stated above, be distinct.

A primary *psychological* predictor of minimisation and reuse intentions is an 'Active Concern' for waste issues. As recommended for recycling, a publicity campaign, using shock tactics and emphasising personal environmental responsibility is needed. However, this must be aimed specifically at these behaviours. One major finding of this research is that the determinants of all three behaviours are different. Hence, publicity campaigns must play both generally on the waste problem, but most crucially deal with only one behaviour at a time and emphasise how that can help. Mixing the behaviours will ensure that only the most basic of messages gets through, leading to little or no intention change. Indeed, emphasis must be placed on the fact that it's far better to reduce waste than produce it and then recycle it. This is vital.

There is more scope with minimisation and reuse intentions, however, to emphasise the need for action and personal responsibility. Figures 7.1 and 7.2 clearly show that perception of a 'Threat' from poor waste management is a crucial predictor of a willingness to minimise and reuse, as is the 'Citizenship' factor, emphasising rights and responsibilities towards the environment. Hence, since minimisation and reuse behaviours are so much less developed than recycling behaviour, and also since they are neither normatively nor contextually based, it is recommended that a much larger effort be put into these campaigns on awareness and responsibility. It must be borne in mind at all times that the intention and behaviour base is much lower than for recycling.

Minimisation behaviour Recourse to Figure 7.1 demonstrates that there are a large number of intervening variables between the intention of individuals to minimisation and actual behaviour. Both sets of environmental values are important, and should be dealt with as outlined above. Four situational variables the have an impact, with the most important appearing to be age. As described in explanation of this diagram in Chapter Seven, for every movement up the age scale used in this study there is a large movement in the standard deviation place of the dependent variable, behaviour. Essentially, those in higher age groups reduce their waste more. Linked to this is another situational variable, gender. Females appear to minimise a lot more than males. Hence, a key policy recommendation is that the awareness campaign outlined above should be focused, where possible, at young male audiences. This may not be possible in the sense that one cannot simply send information to young males. However, the messages can be tailored, specifically on minimisation, to that audience by appropriate marketing and communications techniques.

The other two situational variables that appear to be important are 'Policy knowledge' and 'Knowledge sources'. This demonstrates that those who minimise more tend to be more aware of Local Agenda 21, Sustainable Development and use a diverse knowledge base to get information about waste issues. As discussed above in relation to environmental values above, knowledge of LA21 and Sustainable Development should filter through slowly and via LA21 campaigns by local authorities. However, in terms of using more knowledge bases for information about waste, this should be achieved by the awareness campaigns outlined above. The use of various media should therefore broaden the knowledge base people receive and look for information about waste.

One psychological variable had a direct effect on minimisation behaviour. The 'Community/Democracy' scale, measuring inclusion in local decision making on environmental issues and community spirit, appears to show that those who perceive a vibrant community and an active democratic framework tend to minimise more. As discussed in Chapter Seven, this is probably a statistical anomaly, but if this group predicts minimisation more, then it justifies a wide ranging awareness campaign to reach all people.

Reuse behaviour Recourse to Figure 7.2 demonstrates that one environmental value variable intervenes to have a direct effect on reuse behaviour, along with three psychological factors. The same sentiments as expressed concerning environmental values and minimisation behaviour can be used as above. However, the three psychological variables are important here.

First, the 'Convenience and Effort' scale is important, and poses somewhat of a problem from a policy point of view. Making recycling more convenient is quite simple. However, because reuse is not a structured or municipally controlled behaviour, making it more simple and convenient is not as simple. It is recommended that within the campaigns on awareness disseminated to the population and those with kerbside recycling bins, there be the inclusion of advice on how to reuse (e.g. jars for jam storage, tubs to store cold meat in, bottles as vases, etc) and where reused material (e.g. clothes, furniture, bags, etc) can be taken.

The second psychological factor is the 'Motivation to respond' scale and emphasises the response efficacy of acting. This should hopefully be brought out in the awareness campaigns, but special emphasis on the value of reusing should be placed within the reuse campaign.

Finally, those in community groups appear to reuse more ('Total Group Membership' in Figure 7.2). As above, this justifies making the campaigns as wide as possible, so to reach all those in the city, not just those in community, environmental or political groups.

Policy Recommendations: Summary

Table 8.2 below shows the policy recommendations for the three behaviours. As can be seen from the table, it is necessary that the three behaviours have separate campaigns associated with them, since it is believed that the message will penetrate more effectively, especially as this research has found such vast differences in the predictors of each behaviour. Hence, the timing of the campaigns is especially important. It is recommended here that the general campaigns be of short (one month) duration, so as to make a snappy impact. A break of at least one month before the next one should be left. Similarly, literature sent out in the 'Reduce Too' and 'Reuse Too' campaigns should be done at different time intervals.

Table 8.2 Policy recommendations

Behaviour	Policy/Campaign	Focus	Instruments/Actions*
Recycling	*Recycle* Campaign	General public: The waste problem if we don't recycle The need for all to recycle The positive effects of recycling	TV, Radio, Press, Posters, Council Literature, Leaflets**
	Recycle Easy Campaign	Where and how to recycle	Council Tax Bills, Posters, Leaflets**
	Uniformity in recycling sites	Minimum of three 'core' recyclables per site	Adjustment of sites**
	Recycling site changes	All supermarkets have all recyclables	Supermarket - LA negotiation*
	Kerbside bins	Increase as practicable	LA negotiation with collection company**
	Glass collection	For all residents/proportion who need it most	Investigate*
	Communal recycling	For areas of terracing/flats	Investigate*
	Block Leaders	For areas of low recycling	Investigate*
Minimisation	*Help Reduce Waste* Campaign	General public, but messages for young men especially: The threat of waste if we don't reduce it The responsibility of everyone to reduce waste	TV, Radio, Press, Posters, Council Literature, Leaflets**
	Reduce Too Campaign	Those with kerbside bin: The need to reduce waste The benefits of reducing waste How to reduce waste	Targeted address leaflets from LA**
	Local Agenda 21	General public: Importance of environmental sustainability	Strengthen existing campaigns**
Reuse	*Help Reuse Waste* Campaign	General public: The threat of waste if we don't reuse it The responsibility of everyone to reuse waste The positive effects of reuse How and where to reuse	TV, Radio, Press, Posters, Council Literature, Leaflets**
	Reuse Too Campaign	Those with kerbside bin: The need to reuse waste The benefits of reusing waste	Targeted address leaflets from LA**
	Local Agenda 21	General public: Importance of environmental sustainability	Strengthen existing campaigns**

* Local Authority should investigate the possibility of this action
** Local Authority should seriously consider this action

9 Conclusion: household waste in social perspective

The Importance of Individuals in Managing Household Waste

There can be no doubt that the householder possesses the key to sustainable management of municipal waste. The introduction to this text examined a number of alternative approaches that have been used to both reduce waste and divert what waste is produced from landfill. These have had little effect. The evidence presented in Chapters Two to Eight and, in particular, Chapter Seven, provide clear support for the notion that waste minimisation, reuse and recycling rates vary substantially between individual households and for a variety of reasons. The contention in the preface to this book was that economic instruments for resolving environmental dilemmas are outdated. Public acceptance of new fiscal initiatives is low and the rate of environmental degradation is now exceeding the pace of the positive effects of artificial economic adjustments. These facts lead to the compelling conclusion that only changes in individual attitudes towards the environment and with that the consequent shifts in personal behaviour will have the lasting and profound impact on the environment that is so readily needed.

This is not a return to the catastrophist 'Limits to Growth' argument of the early 1970s, but rather a realisation that there are at least desirable or aesthetic, if not natural, limits to the human exploitation of the environment. We are by no means teetering on the brink of the global environmental abyss. However, the occasional freak storm or prolonged drought does bring into focus the likely consequences of global warming. Similarly, the paucity of sites now aesthetically or physically suitable for dumping waste remind us that landfill's game is almost up.

It is perhaps the distant nature of the environmental abyss that is the greatest problem for advocates of a social solution to environmental ills. Because the disaster is always a few generations away, humans can seek ways to maintain their exuberant lifestyles by looking to technology or economics to resolve these issues. A good example of this blinkered view of environmental protection can be seen in California where the deregulation of the electricity industry has led to a classic short-term response. In the face of 'rolling blackouts' by electricity companies that could no longer afford the soaring price of electricity from the power producers, the advice to consumers was to 'sit tight' and wait for the problem to pass. This 'conserve now and enjoy later' approach ignored the fact that rolling blackouts occurred because of the soaring demand for electricity which is increasing all the time. No one defined that overall, long lasting reductions in household use of appliances and other items would mean lower demand and consequently less of need to find new sources of electricity, the subject of President George W. Bush's strategy to take a hard line on energy production in future.

Yet a social solution to environmental problems need not mean a radical reduction in quality of life (which itself is subjective and normatively defined). The research in the preceding chapters has shown that with regard to waste, small changes in lifestyle can result in significant alterations in behaviour. This need not imply recycling everything all of the time. However, recycling some things some of the time is in a sense a huge change in attitudes and behaviour for most individuals who never did so before. The research here shows that beahvioural experience can have such a positive feedback as to affect other environmental actions.

This 'small change' message is used by a number of public agencies and environmental charities to change people's attitudes towards their everyday activities. The 'Are You Doing Your Bit?' Campaign stresses the importance of small actions such as switching off lights and turning off taps to save energy and water, respectively. Similarly, the charity 'Global Action Plan' emphasises the significant impacts of five key environmental behaviours.

What these agencies and others like them recognise is that the technocentric arguments of the 1980s under Thatcherism have been overtaken by events and that it is individual people who now hold the key to creating a sustainable environment. They also recognise what many governments and economists do not, namely that sustainable development is a process, not a step change. We are embarked on a process that will take generations to yield majoritarian attitude change, similar to attitudes towards smoking or drink driving. Yet the results from such change will be consistent and lasting attitudes towards the environment, rather than sporadic and short-term results that are achieved by technological advancement or economic adjustment, which, as De Young (1986) has so compellingly argued, lead to a net reduction in positive attitudes towards environmental protection.

The Social Psychological Approach: Thoughts on the UK Experience

Having presented the argument for a social understanding of environmental problems, there is the issue of how to change the attitudes and behaviours of households towards environmental action. This book has presented a lengthy analysis of how such an approach, based on constructing a flexible framework of behaviour, can be used to plan, implement, analyse and interpret a given study of environmental behaviour. There are two points that require elaboration on the use of such an approach within the UK. First, there is a need to examine the efficacy of this predominantly North American approach within the British context. Second, the place of such research in the wider environmental action debate within the UK needs to be examined.

In terms of the efficacy of this approach within a UK context, it has been shown that those techniques widely applied elsewhere in the world are suitable within a British setting. Although the literature is mostly from the United States and Canada, the variables that have been identified by researchers in those countries have, by and large, been important in influencing behaviour in the current research. As was demonstrated in Chapter Eight, the comparison of the Exeter study to this research has revealed some important differences in how the data are grouped and what factors emerge. However, the approach taken here and in other similar research projects, is that this flexibility is important. Where, however, the Exeter study did make advances was with regard to minimisation and reuse behaviour. It was shown in Chapter Seven that although factors were identified that adequately explained some of these behaviours, the explanation offered by these regression models was significantly lower than that for recycling behaviour. This points to the conclusion that since most American research projects have been interested in examining the determinants of recycling, this is a behaviour that is best understood by researchers. Consequently, waste reduction and reuse are less well defined and poorly understood. The social psychological approach has offered an insight into the determinants of minimisation and reuse behaviour, but has not provided anywhere near conclusive evidence of their explanatory framework. This is not to state that the social psychological approach is flawed, but rather that more exploratory and experimental work is required in order to better understand these more marginal behaviours.

Nonetheless, the approach has proved itself technically in terms of the useful data that are produced and that can be applied to advance both theory and policy, as was demonstrated in Chapter Eight and as will be emphasised later in this chapter. The insights that are offered by the psychologist's use of statistical inference and their more flexible approach to data assumptions permits powerful analyses of data sets with, of course, the 'health warning' that should always accompany statistical inference. Indeed, it has not been the aim of this book to state that statistics are the answer to all human problems, nor that human behaviour is accurately predictable. Rather, statistics are used in this context to show trends in large data sets that relate to rigorous and well-tested theories of human action towards the environment. It is argued here that this is the only way that policy can be adequately formulated and theory advanced.

This is not to dismiss the arguments of qualitative researchers in the UK. This brings us to the second point to be emphasised when considering the social psychological approach. The methodologies advanced by qualitative researchers in the UK have been debated in both Chapters One and Five of this text and have been found wanting in a number of ways. Whilst there is no logical argument against the notion that all human behaviour is unique and that individualist experiences have to be examined over generalisations of pro-environmental behaviour, it is contended here that such a stance provides a position that is self-defeating. Although analysis of human behaviour should take into account as many factors as possible, any meaningful results cannot be obtained from a purely qualitative standpoint. This study has shown that there are certain key causative trends that underlie behaviour that are generalisable and that these can be used to progress our theories about how humans act and how we might positively change behaviour.

Yet hearing the voices of those individuals in society and their perceptions of environmentalism is important. What the statistical tools outlined in Chapter Seven cannot do is examine human decision making within a temporal framework. In other words, whilst we can state that humans act in a given way and that this is likely to be explained by, for example, intrinsic motivation, we do not know why some people are intrinsically motivated and why some people change their motivations for acting. All the quantitative data can tell us is that these factors are important. Indeed, qualitative data provide us with valuable information on the rhetorics of environmentalism and the intricate local contexts underlying behaviour.

The ultimate aim of our research should be to integrate these two approaches that have for too long been juxtaposed. Qualitative researchers need to recognise that quantitative data can offer insights into environmental behaviour that are both theoretically sound and practically essential. Similarly, quantitative researchers need to recognise the invaluable data that can be gained about the process of decision-making and the intricacies of action at the local level. It is argued that any such integrated approach could offer a far more rounded view of environmental behaviour, with the flow of information being circular rather than linear. For instance, qualitative data can be readily used to inform quantitative questionnaire design, whilst quantitative generalisation can be used to form a basis for posing questions regarding individual behaviour change.

This call for unity may or may not be heeded. We are still in the position v˚ere quantitative is equated, at least within geography, with the unacceptable face of social science, where the over-generalisations and statistical excesses of the 1960s and 1970s are often cited as a reason not to use statistical methods. Yet other disciplines have moved on from this argument and have developed methods that provide reliable data for the quantitative researcher. Geography should move beyond the quantitative-qualitative debate and see that as the logical positivists of the 1960s had their faults, so did and do the zealous postmodernists and poststructuralists of today's geography. This book provides a point for debate in itself. It is unashamedly quantitative in its outlook, even though the Exeter study had a large qualitative component. However, the purpose of this book is to show that there is an acceptable face of quantitative geography that seeks not to supercede the Cultural Revolution of the 1990s, but to re-establish common sense and an atmosphere of integration not separation between science and subjectivity.

The Way Forward for Theory and Policy

This book and the research within it has implications for both the theoretical advancement of our understanding of human behaviour as well as our practical understanding of how we can shape action.

Theoretical Implications

From the theoretical standpoint there is a need to introduce two essential ingredients. First, there is a need for holism within the debate on environmental action. Too often studies are published that are either focused upon a particular psychological model or consider only the efficacy of a small number of specific variables with which the researcher is interested. This text has demonstrated the need to move beyond the partisan view of either theoretical alignment or personal interest and examine the holistic view of environmental behaviour. This is not to state that the framework presented here is complete or even the best that can be achieved, but rather to suggest that there is a need to have a unifying mechanism whereby all previous research can be included and analysed easily. Having said this, it is argued here that the framework developed from the literature review and amended in Figure 8.1 does present a compelling organisation of factors that should be used at the very least for debate.

Second, and related to the above point, is the need for the Theory of Reasoned Action (TRA) to provide the basis for our thinking on environmental behaviour. The research presented here demonstrates the efficacy of the central element of the TRA – the intention-behaviour relationship. What has also been termed the 'value-action' gap by qualitative researchers, refers to a fundamental relationship that deserves attention. The disparity between aspirations and reality is a divide that provides probably the best chance of examining the policies that could change behaviour. The amended TRA developed here provides just such a framework for this analysis.

Policy Implications

The applied nature of the research presented within this book presents many exciting opportunities for informing policy with regard to environmental behaviour. Chapter Eight described the detailed policy recommendations that were provided as part of the Exeter study. These recommendations demonstrate the useful practical measures that can be derived from such a study. Overall, however, the implications for policy are based around the need to target measures at specific elements of the population, which have behavioural specificity implicit within them.

The data presented in the preceding chapters and from other studies shows the need to treat different behaviours with care. They have different determinants that require alternative policies. Hence, treating 'waste' as one behavioural entity is no longer acceptable. Policy makers need to investigate the underlying factors of each behavioural category and proceed on that basis. Indeed, policies that are implemented need to be based on targeted messages at identified sections of the population. In this way, messages are not wasted and resources saved. Of course, the crucial factor here is research that can answer these preliminary questions. This in itself is a question of resources which only government can answer, but until we know what people do and why, we cannot make policies that will tangibly change their behaviour.

Prospect

This book provides fuel for the debate on the social solutions available for resolving environmental problems, as well as a concise argument on how this can be achieved. We are still at the stage of contemplation and, in terms of environmental change, we are still blissfully unappreciative of what problems could face us in the future if we do not act now. But action must come from the bottom upward; from individuals through society. This approach, recognising the valuable contribution each individual has to make to environmental preservation and sustainability, will enable environmental behaviour to become socially normative within several generations. Yet an approach based on economic adjustment or technological 'patching-up' will prove fruitless.

This text has provided a salient example of how to examine human attitudes and behaviours towards one environmental activity. Yet the approach can be extrapolated to many other behaviours. Such a social-psychological perspective provides an opportunity to advance both theory and policy. With the integration of qualitative methodologies, the scope widens still. We are still a long way off providing a 'theory' of environmental behaviour; we may never do so. But what is clear is that every one of us, by our actions every day, contribute to the future environmental state of our planet, be that a future of sustainability or continued degradation.

Bibliography

Aberg, H., Dahlman, S., Shanahan, H. and Saljo, R. (1996), 'Towards sound environmental behavior: exploring household participation in waste management', *Journal of Consumer Policy*, vol.19 pp.45-67.

Agyeman, J. and Evans, B. (1995), 'Sustainability and Democracy: community participation in Local Agenda 21', *Local Government Policy Making*, vol.22 (2), pp.35-40.

Ajzen, I. (1991), 'The Theory of Planned Behavior', *Organizational Behavior and Human Decision Processes*, vol.50 pp.179-211.

Ajzen, I. and Madden, T.J. (1986), 'Prediction of goal-directed behavior: attitudes, intentions, and perceived behavioral control', *Journal of Experimental Social Psychology*, vol.22 pp.453-474.

Albrecht, D., Bultena, G., Hoiberg, E. and Nowak, P. (1982), 'The New Environmental Paradigm scale', *Journal of Environmental Education*, vol.13 pp.39-43.

Arbutnot, J. (1977), 'The roles of attitudinal and personality variables in the prediction of environmental behavior and knowledge', *Environment and Behavior*, vol.9 (2), pp.217-232.

Arcury, T. A., Johnson, T. P. and Scollay, S. J. (1986), 'Ecological worldview and environmental knowledge: the "New Environmental Paradigm"', *Journal of Environmental Education*, vol.17 pp.35-40.

Axelrod, L.J. and Lehman, D.R. (1993), 'Responding to environmental concerns: what factors guide individual actions', *Journal of Environmental Psychology*, vol.13 (2), pp.149-160.

Bagozzi, R.P. (1992), 'The self-regulation of attitudes, intentions and behavior', *Social Psychology Quarterly*, vol.55 (2), pp.178-204.

Bailey, I. (1999), 'Flexibility, harmonization and the single market in EU environmental policy: the Packaging Waste Directive', *Journal of Common Market Studies*, vol.37 (4), pp.549-571.

Baldassare, M. and Katz, C. (1992), 'The personal threat of environmental problems as predictor of environmental practices', *Environment and Behavior*, vol.24 (5), pp.602-616.

Ball, R. and Lawson, S.M. (1990), 'Public attitudes towards glass recycling in Scotland', *Waste Management and Research*, vol.8 pp.177-182.

Barr, S. (1998), *Making Agenda 21 work: recycling use in Oxfordshire*, Unpublished undergraduate dissertation Department of Geography, University of Exeter.

Barr, S. (2001), *Factors Influencing Household Attitudes and Behaviours Towards Waste Management in Exeter, Devon*, Unpublished PhD thesis, Department of Geography, University of Exeter.

Becker, L.J., Seligman, C., Fazio, R.H. and Darley, J.M. (1981), 'Relating attitudes to residential energy use', *Environment and Behavior*, vol.13 (5), pp.590-609.

Bell, J., Fisher, J.D., Baum, A. and Greene, T.E. (1990), *Environmental Psychology*, Holt, Rinehart and Winston, Fort Worth.

Bell, M. and Evans, D. (1997), 'Greening the "Heart of England": redemptive science, citizenship and "symbol of hope for the nation"', *Environment and Planning D: Society and Space*, vol.15 (3), pp.257-279.

Berger, I.E. (1997), 'The demographics of recycling and the structure of environmental behavior', *Environment and Behavior*, vol.29 (4), pp.515-531.

Blake, D.E., Guppy, N. and Urmetzer, P. (1997), 'Canadian public opinion and environmental action: evidence from British Columbia', *Canadian Journal of Political Science*, vol.30 (3), pp.451-472.

Blamey, R. (1998), 'The activation of moral norms: extending Schwartz's model', *Environment and Behavior*, vol.30 (5), pp.676-708.

Blocker, T.J. and Eckberg, D.L. (1997), 'Gender and environmentalism: results from the 1993 General Social Survey', *Social Science Quarterly*, vol.78 (4), pp.841-858.

Blowers, A. (1993), 'The time for change', in Blowers, A. (ed), *Planning for a sustainable environment*, Earthscan, London, pp.1-18.

Boldero, J. (1995), 'The prediction of household recycling of newspapers: the role of attitudes, intentions, and situational factors', *Journal of Applied Social Psychology*, vol.25 (5), pp.440-462.

Bord, R.J. and O'Conner, R.E. (1997), 'The gender gap in environmental attitudes: the case of perceived vulnerability to risk', *Social Science Quarterly*, vol.78 (4), pp.830-840.

Bryce, W.J., Day, R. and Olney, T.J. (1997), 'Commitment approach to motivating community recycling: New Zealand curbside trial', *The Journal of Consumer Affairs*, vol.31 (1), pp.27-52.

Bryman, A. and Cramer, D. (1996), *Quantitative Data Analysis with MINITAB*, Routledge, London.

Burgess, J., Harrison, C.M. and Filius, P. (1998), 'Environmental communication and the cultural politics of environmental citizenship', *Environment and Planning A*, vol.30 (8), pp.1445-1460.

Burn, S.M. (1991), 'Social psychology and the stimulation of recycling behaviors: the Block Leader approach', *Journal of Applied Social Psychology*, vol.21 (8), pp.611-629.

Burn, S.M. and Oskamp, S. (1986), 'Increasing community recycling behavior with persuasive communication and public commitment', *Journal of Applied Social Psychology*, vol.16 (1), pp.29-41.

Buttel, F.H. (1987), 'New directions in environmental sociology', *American Review of Sociology*, vol.13 pp.465-488.

Buttel, F.H. and Flinn, W.L. (1978), 'The politics of environmental concern. The impacts of party identification and political ideology on environmental attitudes', *Environment and Behavior*, vol.10 (1), pp.17-36.

Campbell, C. (1999), 'The Easternisation of the West' in Wilson, B and Cresswell, J. (eds), *New Religious Movements: Challenge and Response*, Routledge, London, pp.35-48.

Canter, D. and Donald, I. (1987), 'Environmental psychology in the United Kingdom' in Stokols D. and Atman, I. (eds), *Handbook of Environmental Psychology*, Wiley, New York, pp.1281-1310.

Carson, R. (1962), *Silent Spring*, Houghton Miflin, Boston.

Catton, W.R. and Dunlap, R.E. (1978), 'Environmental sociology: a new paradigm', *The American Sociologist*, vol.13 pp.41-49.

Central Statistical Office (1995, 1996, 1997, and 1998), *Social Trends*, HMSO, London.

Chan, K. (1998), 'Mass communication and proenvironmental behaviour: waste recycling in Hong Kong', *Journal of Environmental Management*, vol.52 pp.317-325.

Christie, I. and Jarvis, L. (1999), 'Rural spaces and urban jams' in Jowell, R., Curtice, J. Park, A. and Thomson, K. (eds), *British Social Attitudes: the 16th report. Who shares New Labour values?*, Ashgate, Aldershot, pp. 113-134.

Clayton, K. (1976), 'Environmental sciences/studies: a decade of attempts to discover a curriculum', *Area*, vol.8 (2), pp.98-101.

Cloke, P., Philo, C. and Sadler, D. (1991), *Approaching Human Geography: an introduction to contemporary theoretical debates*, Paul Chapman Publishing, London.

Commission of the European Communities (CEC), (1992), *Towards sustainability; A European Community programme of policy and action in relation to the environment and sustainable development*, EC Fifth Environmental Action Programme, Cm (92), 23/11 Final. CEC, Brussels.

Corral-Verdugo, V. (1997), 'Dual "realities" of conservation behavior: self reports Vs observations of re-use and recycling behavior' *Journal of Environmental Psychology*, vol.17 pp.135-146.

Corral-Verdugo, V., Bernache, G., Encinas, L. and Garibaldi, L. (1994-1995), 'A comparison of two measures of reuse and recycling behaviour: self-report and material culture', *Journal of Environmental Systems*, vol.23 (4), pp.313-327.

Dalton, R. and Rohrschneider, R. (1998), 'The greening of Europe', in Jowell, R., Curtice, J., Park, A., Brook, L., Thomson, K. and Bryson, C. (eds), *British and European Social Attitudes*, Ashgate, Aldershot, pp.101-124.

Daneshvary, N., Daneshvary, R. and Schwer, R. K. (1998), 'Solid-waste recycling behavior and support for curbside textile recycling', *Environment and Behaviour*, vol.30 (2), pp.144-161.

Dawson, J.A. and Doornkamp, J.C. (1973), 'Introduction', in Dawson, J.A., and Doornkamp, J.C. (eds), *Evaluating the Human Environment: Essays in Applied Geography*, Edward Arnold, London, pp.1-3.

Department of the Environment (1995), *Making waste work: A strategy for sustainable waste management in England and Wales*, HMSO, London.

Department of the Environment, Transport and the Regions (1998), *Less Waste: More Value. Consultation Paper on a Waste Strategy for England and Wales*, DETR, London.

Department of the Environment, Transport and the Regions (1999a), *A way with waste: A draft waste strategy for England and Wales*, DETR, London.

Department of the Environment, Transport and the Regions (1999b), *Limiting Landfill: A consultation paper on limiting landfill to meet the EC Landfill Directive's targets for the landfill of biodegradable municipal waste*, DETR, London.

Department of the Environment, Transport and the Regions (1999c), *A better quality of life: A strategy for Sustainable Development for the UK*, DETR London.

Department of the Environment, Transport and the Regions (2000), *Waste Strategy 2000*, The Stationary Office, London.

Derksen, L. and Gartell, J. (1993), 'The social context of recycling', *American Sociological Review*, vol.58 (3), pp.434-442.

De Young, R. (1985-1986), 'Encouraging environmentally appropriate behaviour: the role of intrinsic motivation', *Journal of Environmental Systems*, vol.15 (4), pp.281-292.

De Young, R. (1986), 'Some psychological aspects of recycling', *Environment and Behavior*, vol.18 (4), pp.435-449.

De Young, R. (1988-1989), 'Exploring the difference between recyclers and non-recyclers: the role of information', *Journal of Environmental Systems*, vol.18 (4), pp.341-351.

De Young, R. (1990), 'Recycling as appropriate behaviour: a review of survey data from selected recycling education programmes in Michigan', *Resources, Conservation, Recycling*, vol.3 pp.253-266.

De Young, R. and Kaplan, S. (1985-1985), 'Conservation behaviour and the structure of satisfactions', *Journal of Environmental Systems*, vol.15 (3), pp.233-242.

Diamond, W. D. and Leowy, B. Z. (1991), 'Effects of probabilistic rewards on recycling attitudes and behavior', *Journal of Applied Social Psychology*, vol.21 (19), pp.1590-1607.

Diekman, A. and Preisendorfer, P. (1998), 'Environmental behavior: discrepancies between aspirations and reality', *Rationality and Society*, vol.10 (1), pp.79-102.

Dietz, T., Stern, P. C. and Guagnano, G. A. (1998), 'Social structural and social psychological bases of environmental concern', *Environment and Behavior*, vol.30 (4), pp.450-471.

Dillman, R. A. (1978), *Mail and Telephone Surveys*, Wiley, New York.

Dunlap, R. E. (1975), 'The impact of political orientation on environmental attitudes and actions', *Environment and Behavior*, vol.7 (4), pp.428-453.

Dunlap, R. E. and Mertig, A. G. (1995), 'Global concern for the environment: is affluence a prerequisite?', *Journal of Social Issues*, vol.51 (4), pp.121-137.

Dunlap, R. E. and Van Liere, K. D. (1978), 'The "New Environmental Paradigm"', *Journal of Environmental Education*, vol.9 pp.10-19.

Eden, S. (1993), 'Individual environmental responsibility and its role in public environmentalism', *Environment and Planning A*, vol.25 pp.1743-1758.

Ehrlich, H. J. (1969), 'Attitudes, behaviour and the intervening variables', *American Sociologist*, vol.4 (1), pp.29-34.

Eiser, J. R. (1986), *Social Psychology: Attitudes, Cognition and Social Behaviour*, Cambridge University Press, Cambridge.

Etzioni, A. (1993), *The Spirit of Community: Rights, Responsibilities, and the Communitarian Agenda*, Fontana/Harper-Collins, London.

Etzioni, A. (1995a), 'Introduction', in Etzioni, A. (ed.), *New Communitarian Thinking: Persons, Virtues, Institutions, and Communities*, University of Virginia Press, London, pp.1-15.

Etzioni, A. (1995b), 'Old chestnuts and new spurs', *ibid.* pp.16-34.

Evans, B. (1995), 'Local environmental policy and Local Agenda 21', *Area*, vol.27 (2), pp.163-164.

Everett, J.W. and Pierce, J.J. (1991-1992), 'Social networks, socioeconomic status and environmental collective action: residential curbside Block Leader recycling', *Journal of Environmental Systems*, vol.21 (1), pp.65-84.

Exeter City Council (1997), *A Local Agenda 21 for Exeter*, Exeter City Council, Exeter.

Exeter City Council (2000a), *Exeter City Council: key aims and objectives 2000-2001*, Exeter City Council, Exeter, www.exeter.gov.uk.

Exeter City Council (2000b), *A Waste Strategy for Exeter*, Exeter City Council, Exeter, Via www.exeter.gov.uk.

Fazio, R.H. and Zanna, M.P. (1978), 'Attitude qualities relating to the strength of the attitude-behavior relationship', *Journal of Experimental Social Psychology*, vol.14 (4), pp.398-408.

Fishbein, M. (1967), 'Attitude and the prediction of behaviour', in Fishbein, M. (ed), *Readings in attitude theory and measurement*, Wiley, New York, pp.477-492.

Fishbein, M. and Ajzen, I. (1975), *Belief, Attitude, Intention and Behavior: An Introduction to Theory and Research*, Addison-Wesley, Reading, MA.

Folz, D. H. (1991), 'Recycling program design, management, and participation: a national survey of municipal experience', *Public Administration Review*, vol.51 (3), pp.222-231.

Fowler, F. J. JR. (1988), *Survey Research Methods*, SAGE, Thousand Oaks, Ca.

Fowler, F. J. JR. (1995), *Improving Survey Questions; Design and Evaluation*, SAGE, Thousand Oaks, Ca.

Freeman, C., Littlewood, S. and Whitney, D. (1996), 'Local government and the emerging models of participation in the Local Agenda 21 process', *Journal of Environmental Planning and Management*, 39 (1), pp.65-78.

Gamba, R. J. and Oskamp, S. (1994), 'Factors influencing community residents' participation in commingled curbside recycling programs', *Environment and Behavior*, vol.26 (5), pp.587-612.

Gardner, G. (1976), *Social Surveys for Social Planners*, OU Press, Milton Keynes.

Gatrsleben, B. and Vlek, Ch. (1998), 'Household consumption, quality of life and environmental impacts' in Noorman, K.J. and Uiterkamp, A.J.M. (eds), *Green Households? Domestic Consumers, Environment and Sustainability*, Earthscan, London, pp.141-183.

Geller, E. S. (1989), 'Applied Behavior Analysis and social marketing: an integration for environmental preservation', *Journal of Social Issues*, vol.45 (1), pp.17-36.

Geller, E. S., Winett, R. and Everett, P. B. (1982), *Preserving the Environment: New Strategies for Behavior Change*, Pergamon, New York.

Gibbs, D., Longhurst, J. and Braithwaite, C. (1996), 'Moving towards sustainable development? Integrating economic development and the environment in local authorities', *Journal of Environmental Planning and Management*, vol.39 (3), pp.317-332.

Gibbs, D., Longhurst, J. and Braithwaite, C. (1998), '"Struggling with sustainability": weak and strong interpretations of sustainable development within local authority policy', *Environment and Planning A*, vol.30 (8), pp.1351-1365.

Goldenhar, L. M. and Connell, C. M. (1992-1993), 'Understanding and predicating recycling behavior; an application of the theory of reasoned action', *Journal of Environmental Systems*, vol.22 (1), pp.91-103.

Gray, S. (1971), *The Electoral Register: Practical Information for use when Drawing Samples, both for Interview and Postal Survey*, OPCS, London.

Grob, A. (1995), 'A structural model of environmental attitudes and behaviour', *Journal of Environmental Psychology*, vol.15 pp.209-220.

Guagnano, G.A., Ditez, T. and Stern, P.C. (1994), 'Willingness to pay for public goods: a test of the contribution model', *Psychological Science*, vol.5 (6), pp.411-415.

Guagnano, G.A., Stern, P.C. and Dietz, T. (1995), 'Influences on attitude-behavior relationships: a natural experiment with curbside recycling', *Environment and Behavior*, vol.27 (5), pp.699-718.

Guth, J.L., Green, J.C., Kellstedt, L.A. And Schmidt, C E. (1995), 'Faith and Environment: Religious beliefs and attitudes on environmental policy', *American Journal of Political Science*, vol.39 (2), pp.364-392.

Hallin, P. O. (1995), 'Environmental concern and environmental behaviour in Foley, a small town in Minnesota', *Environment and Behavior*, vol.27 (4), pp.558-578.

Hawthorne, M. and Alabaster, T. (1999), 'Citizen 2000: development of a model of environmental citizenship', *Global Environmental Change*, vol.9 pp.25-43.

Heberlein, T.A. (1989), 'Attitudes and environmental management', *Journal of Social Issues*, vol.45 (1), pp.37-57.

Heberlein, T.A. and Black, J.S. (1981), 'Cognitive consistency and environmental action', *Environment and Behavior*, vol.13 (6), pp.717-734.

Hermandez, O., Rawlons, B. and Schwartz, R. (1999), 'Voluntary recycling in Quito: factors associated with participation in a pilot program', *Environment and Urbanization*, vol.11 (2), 145-159.

Hines, J. M., Hungerford, H. R. and Tomera, A. N. (1987), 'Analysis and synthesis of research on responsible environmental behavior: a meta analysis', *Journal of Environmental Education*, vol.18 (2), pp.1-8.

Hopper, J. R. and Nielsen, J. M. (1991), 'Recycling as altruistic behavior: normative and behavioral strategies to expand participation in a community recycling programme', *Environment and Behavior*, vol.23 (2), pp.195-220.

Howenstine, E. (1993), 'Market segmentation for recycling', *Environment and Behavior*, vol.25 (1), pp.86-102.

Humphrey, C.R., Bord, R.J., Hammond, M.M. and Mann, S.H. (1977), 'Attitudes and conditions for cooperation in a paper recycling program', *Environment and Behavior*, vol.9 (1), pp.107-124.

Inglehart, R. (1977), *The Silent Revolution: Changing Values and Political Styles Among Western Politics*, Princeton University Press, Princeton, NJ.

Inglehart, R. (1981), 'Post-materialism in the environment of insecurity', *American Political Science Review*, 75 (4), pp.880-900.

Inglehart, R. (1990), *Culture Shift in Advanced Industrial Society*, Princeton University Press, Princeton, NJ.

Jacobs, H.E., Bailey, J.S. and Crews, J.I. (1984), 'Development and analysis of a community-based resource recovery program', *Journal of Applied Behavior Analysis*, vol.17 (2), pp.127-145.

Jones, R.E. (1990), 'Understanding paper recycling in an institutionally supportive setting: an application of the Theory of Reasoned Action', *Journal of Environmental Systems*, vol.19 (3), pp. 307-321.

Jones, R.E. and Dunlap, R.E. (1992), 'The social bases of environmental concern: have they changed over time?', *Rural Sociology*, vol.57 (1), pp.28-47.

Kaiser, F.G. (1998), 'A general measure of ecological behavior', *Journal of Applied Social Psychology*, vol.28 (5), pp.395-422.

Kaiser, F.G., Wolfing, S. and Fuher, U. (1999), 'Environmental attitude and ecological behaviour', *Journal of Environmental Psychology*, vol.19 pp.1-19.

Kallgren, C.A. and Wood, W. (1986), 'Access to attitude-relevant information in memory as a determinant of attitude-behaviour consistency', *Journal of Experimental Social Psychology*, vol.22 (4), pp.328-338.

Kantola, S.J., Syme, G.J. and Campbell, N.A. (1982), 'The role of individual difference and external variables in a test of the sufficiency of Fishbein's model to explain behavioural intentions to conserve water', *Journal of Applied Social Psychology*, vol.12 (1), pp.70-83.

Kearns, A. (1995), 'Active citizenship and local governance: political and geographical dimensions', *Political Geography*, vol.14 (2), pp.155-175.

Kitchin, R.M., Blades, M. and Golledge, R.D. (1997), 'Relations between psychology and geography', *Environment and Behavior*, vol.29 (4), pp.554-573.

Klinebeerg, S.L., McKeever, M. and Rothenbach, B. (1998), 'Demographic predictors of environmental concern: it does make a difference how it's measured', *Social Science Quarterly*, vol.79 (4), 734-753.

Kok, G. and Siero, S. (1985), 'Tin recycling: awareness, comprehension, attitude, intention and behavior', *Journal of Economic Psychology*, vol.6 pp.157-173.

Krause, D. (1993), 'Environmental consciousness: an empirical study', *Environment and Behavior*, vol.25 (1), pp.124-142.

Krueger, R.A. (1994), *Focus Groups: a practical guide for applied research*, SAGE, Thousand Oaks, Ca.

Kuhn, R. and Jackson, E.L. (1989), 'Stability of factor structures in the measurement of public environmental attitudes', *Journal of Environmental Education*, vol.20 (3), pp.27-32.

Lam, S. (1999), 'Predicting intentions to conserve water from the Theory of Planned Behavior, perceived moral obligation, and perceived water right', *Journal of Applied Social Psychology*, vol.29 (5), pp.1058-1071.

Lasana, F.M. (1992), 'Distinguishing potential recyclers from nonrecyclers: a basis for developing recycling strategies', *Journal of Environmental Education*, vol.23 (2), pp.16-23.

Lasana, F. M. (1993), 'A comparative analysis of curbside recycling behaviour in urban and suburban communities' *Professional Geographer*, vol.45 (2), pp.169-179.

Lea, S. (1998), *Statistics and Research Methods: Quantitative Component* and *Advanced Statistics: Multivariate Analysis II: Manifest Variables Analyses*, SPSS and MINITAB tutorials (School of Psychology website, via: www.exeter.ac.uk).

Lindsay, J.J. and Strathman, A. (1997), 'Predictors of recycling behavior: an application of a modified health belief model', *Journal of Applied Social Psychology*, vol.27 (20), pp.1799-1823.

Liska, A.E. (1984), 'A critical examination of the causal structure of the Fishhbein/Ajzen attitude-behavior model', *Social Psychology Quarterly*, vol.47 (1), pp.61-74.

Littlewood, S. and While, A. (1997), 'A new agenda for governance? Agenda 21 and the prospects for holistic decision making', *Local Government Studies*, vol.23 (4), pp.110-123.

Ludwig, T.D., Gray, T.W. and Rowell, A. (1998), 'Increasing recycling in academic buildings: a systematic replication', *Journal of Applied Behavior Analysis*, vol.31, pp.683-686.

Luyben P.D. (1982), 'Prompting thermostat setting behavior: public response to a Presidential appeal for conservation', *Environment and Behavior*, vol.14 (1), pp.113-128.

Luyben, P.D. and Bailey, J.S. (1979), 'Newspaper recycling: the effects of rewards and proximity of containers', *Environment and Behavior*, vol.11 (4), pp.539-557.

Macnaghten, P. and Jacobs, M. (1997), 'Public identification with sustainable development: investigating cultural barriers to participation', *Global Environmental Change*, vol.7 (1), pp.5-24.

Macnaghten, P. and Urry, J. (1998), *Contested Natures*, SAGE, London.

Macey, S.M. and Brown, M.A. (1983), 'Residential energy conservation: the role of past experience in repetitive household behavior', *Environment and Behavior*, vol.15 (2), pp.123-141.

Mainieri, T, Barnett, E.G., Valdero, T.R., Unipan, J.B. and Oskamp, S. (1997), 'Green buying: the influence of environmental concern on consumer behavior', *The Journal of Social Psychology*, vol.137 (2), pp.189-204.

Maloney, M.P. and Ward, M.P. (1973), 'Ecology: let's hear it from the people. An objective scale for the measurement of ecological attitude and knowledge', *American Psychologist*, vol.28 pp.583-586.

Maloney, M.P., Ward, M.P. and Braucht, G.N. (1975), 'A revised scale for the measurement of ecological attitudes and knowledge', *American Psychologist*, vol.30 pp.787-790.

Martinez, M.D. and Scicchitano, M.J. (1998), 'Who listens to trash talk? Education and public media effects in recycling behavior', *Social Science Quarterly*, vol.79 (2), pp.287-300.

May, T. (1997), *Social Science Research: Issues and Methods*, OU Press, Buckingham.

McCarty, J.A. and Shrum, L.J. (1994), 'The recycling of solid wastes: personal values, value orientations and attitudes about recycling as antecedents of recycling behavior', *Journal of Business Research*, vol.30 pp.53-62.

McCaul, K.D. and Kopp, J.T. (1984), 'Effects of goal setting and commitment on increasing metal recycling', *Journal of Applied Psychology*, vol.67 pp.377-379.

McDonald, S. and Ball, R. (1998), 'Public participation in plastics recycling schemes', *Resources, Conservation and Recycling*, vol.22 pp.123-141.

McGuinness, J., Jones, A.P. and Cole, S.G. (1977), 'Attitude correlates of recycling behaviour', *Journal of Applied Psychology*, vol.62 (4), pp.376-384.

McKenzie-Mohr, D. and Oskamp, S. (1995), 'Psychology and sustainability: an introduction', *Journal of Social Issues*, vol.51 (4), pp.1-14.

McKenzie-Mohr, D., Nemiroff, L.S., Beers, L. and Desmarais, S. (1995), 'Determinants of responsible environmental behavior', *Journal of Social Issues*, vol.51 (4), pp.139-156.

Meadows, D., Randers, J. and Berhens III, W. (1972), *The Limits to Growth*, Universe Books, New York.

Minton, A.P. and Rose, R.L. (1997), 'The effect of environmental concern on environmentally friendly consumer behavior: an exploratory study', *Journal of Business Research*, vol.40 pp.37-48.

Moore, P. and Cobby, J. (1998), *Introductory Statistics for Environmentalists*, Prentice Hall, London.

Munton, R. (1997), 'Engaging sustainable development: some observations on progress in the UK', *Progress in Human Geography*, vol.21 (2), pp.151-159.

Murphree, D.W., Wright, S.A. and Ebaugh, D.R. (1996), 'Toxic waste siting and community resistance: how co-optation of local citizen opposition failed', *Sociological Perspectives*, vol.39 (4), pp.447-463.

Myers, G. and Macnaghton, P. (1998), 'Rhetorics of environmental sustainability: commonplaces and places', *Environment and Panning A*, vol.30 (2), pp.333-353.

Nazarea, V., Rhoades, R., Bontoyan, E. and Flora, G. (1998), 'Defining indicators which make sense to local people intra-cultural variation in perceptions of natural resources', *Human Organization*, vol.57 (2), pp.157-170.

Neuman, W.C. (1994), *Social Research Methods: Qualitative and Quantitative Approaches*, Allyn and Bacon, Needham Heights, Ma.

Newby, H. (1996), 'Citizenship in a green world: global commons and human stewardship', in Bulmer, M. and Rees, A (eds), *Citizenship Today*, UCL Press, London, pp.209-221.

Noorman, K.J., Biesiot, W and Uiterkamp, A.J.M. (1998), 'Household metabolism in the context of environmental quality', in Noorman, K.J. and Uiterkamp, A.J.M. (eds), *Green*

Households? Domestic Consumers, Environment and Sustainability, Earthscan, London, pp.7-34.

Norris, P. (1997), 'Are we all green now? Public opinion and environmentalism in Britain', *Government and Opposition*, vol.32 (3), pp.320-339.

Office for National Statistics (ONS), (1998), *Family Spending: A report on the 1997-98 Family Expenditure Survey*, The Stationary Office, London.

Office of Population Censuses and Surveys (OPCS), 1992-93), *Census: County Report: Devon* (Parts 1 and 2), HMSO, London.

Olsen, M.E. (1981), 'Consumers' attitudes toward energy conservation', *Journal of Social Issues*, vol.37 (2), pp.108-131.

O'Riordan, T. (1985), 'Future directions in environmental policy', *Environment and Planning A*, vol.17 (11), pp.1431-1446.

O'Riordan, T. (1989), 'The challenge for environmentalism', in Peet, R. and Thrift, N. (eds), *New Models in Geography*, Unwin Hyman, London, pp.77-102.

O'Riordan, T. (1993), 'The politics of sustainability', in Turner, R.K. (ed), *Sustainable Environmental Economics and Management*, Belhaven, London, pp.37-69.

O'Riordan, T. (1997a), 'Sustainability and New Labour radicalism', *ECOS*, vol.18 (1), pp.12-15.

O'Riordan, T. (1997b), 'Labour's greenish credentials', *ECOS*, vol.18 (2), pp.2-5.

Oskamp, S. (1995a), 'Applying social psychology to avoid ecological disaster', *Journal of Social Issues*, vol.51 (4), pp.217-239.

Oskamp, S. (1995b), 'Resource conservation and recycling: behavior and policy', *Journal of Social Issues*, vol.51 (4), pp.157-177.

Oskamp, S., Harrington, M.J., Edwards, T.C., Sherwood, D.L., Okuda, S.M. and Swanson, D.C. (1991), 'Factors influencing household recycling behavior', *Environment and Behavior*, vol.23 (4), pp.494-519.

Painter, J. and Philo, C. (1995), 'Spaces of citizenship: an introduction', *Political Geography*, vol.14 (2), pp.107-120.

Pearce, D. (1993), *Blueprint 3: Measuring Sustainable Development*, Earthscan, London.

Pearce, D., Markandya, A. and Barbier, E. (1989), *Blueprint for a Green Economy*, Earthscan, London.

Pelletier, L.C., Legault, L.R. and Tuson, K.M. (1996), 'The environmental satisfaction scale: a measure of satisfaction with local environmental conditions and government environmental policies', *Environment and Behavior*, vol.28 (1), pp.5-26.

Pelletier, L.G., Green-Demers I., Tuson, K.M., Noels, L. and Beaton, A.M. (1998), 'Why are you doing things for the environment? The Motivation Toward the Environment Scale (MTES)', *Journal of Applied Social Psychology*, vol.28 (5), pp.457-468.

Petts, J. (1995), 'Waste management strategy development: a case study of community involvement and consensus-building in Hampshire', *Journal of Environmental Planning and Management*, vol.38 (4), pp.519-536.

Reid, D.H., Luyben, P.D., Rawers, R.J. nad Bailey, J. (1976), 'Newspaper recycling behaviour: the effects of prompting and proximity of containers', *Environment and Behavior*, vol.8 (3), pp.471-482.

Roberts, J.A. and Bacon, D.R. (1997), 'Exploring the subtle relationships between environmental concern and ecologically conscious consumer behavior', *Journal of Business Research*, vol.40 pp.79-89.

Robinson, G.M. (1998), *Methods and Techniques in Human Geography*, Wiley, Chichester.

Rydin, Y. (1998), *Urban and Environmental Planning in the UK*, Macmillan, Basingstoke.

Sabini, J. (1992), *Social Psychology*, Norton & Co., New York.

Samdahl, D.M. and Robertson, R. (1989), 'Social determinants of environmental concern: specification and test of the model', *Environment and Behavior*, vol.21 (1), pp.57-81.

Schahn, J. and Holzer, E. (1990), 'Studies of individual environmental concern: the role of knowledge, gender and background variables', *Environment and Behavior*, vol.22 (6), pp.767-786.

Schultz, P.W. (1998), 'Changing behavior with normative feedback interventions: a field experiment with curbside recycling', *Basic and Applied Social Psychology*, vol.21 (1), pp.25-36.

Schultz, P.W. and Oskamp, S. (1996), 'Effort as a moderator of the attitude-behavior relationship: general environmental concern and recycling', *Social Psychology Quarterly*, vol.59 (4), pp.375-383.

Schultz, P.W., Oskamp, S. and Mainieri, T. (1995), 'Who recycles and when? A review of personal and situational factors', *Journal of Environmental Psychology*, vol.15 pp.105-121.

Schwartz, S.H. (1977), 'Normative influences on Altruism' in Berkowitz, L. (ed), *Advances in Experimental Social Psychology*, vol.10, Academic Press, New York, pp.221-279.

Scott, D. and Willits, F.K. (1994), 'Environmental attitudes and behavior: a Pennsylvania Survey', *Environment and Behavior*, vol.26 (2), pp.239-260.

Segun, C., Pelletier, L.G. and Hunsley, J. (1998), 'Toward a model of environmental activism', *Environment and Behavior*, vol.30 (5), pp.628-652.

Selman, P. (1994), 'Canada's Environmental Citizens: innovation and partnership for sustainable development', *British Journal of Canadian Studies*, vol.9 (1), 44-52.

Selman, P. (1996), *Local Sustainability: Planning and Managing Ecologically Sound Places*, Chapman, London.

Selman, P. (1998), 'Local Agenda 21: substance or spin?', *Journal of Environmental Planning and Management*, vol.41 (5), pp.533-553.

Selman, P. and Parker, J. (1997), 'Citizenship, civicness and social capital in Local Agenda 21', *Local Environment*, vol.2 (2), pp.171-184.

Shaw, G. and Wheeler, D. (1994), *Statistical Techniques in Geographical Analysis*, Halsted, New York (Second Edition).

Sia, A.P., Hungerford, H.R. and Tomera, A.N. (1985), 'Selected predictors of environmental behavior: an analysis', *Journal of Environmental Education*, vol.17 pp.31-40.

Simmons, D.A. and Widmar, R. (1989-1990), 'Participation in household solid waste reduction activities: the need for public education', *Journal of Environmental Systems*, vol.19 (4), pp.323-330.

Skrentny, J.D. (1993), 'Concern for the environment: a cross-national perspective' International', *Journal of Public Opinion Research*, vol.5 (4), pp.334-352.

Smith, D.M. (1999), 'Geography, community and morality', *Environment and Planning A*, vol. 31 pp.19-35.

Steel, B.S. (1996), 'Thinking globally and acting locally? Environmental attitudes, behaviour and activism', *Journal of Environmental Management*, vol.47 pp.27-36.

Stern, P.C. (1992), 'Psychological dimensions of global environmental change', *Annual Review of Psychology*, vol.43 pp.269-309.

Stern, P.C. and Oskamp, S. (1987), 'Managing scarce environmental resources' in Stokols, D and Altman, I. (eds), *Handbook of Environmental Psychology*, vol.2, Wiley, New York, pp.1043-1088.

Stern, P.C., Dietz, T. and Kalof, L. (1993), 'Value orientations, gender, and environmental concern', *Environment and Behavior*, vol.25 (3), pp.322-348.

Stern, P.C., Dietz, T. and Black, J.S. (1985-1986), 'Support for environmental protection: the role of moral norms', *Population and Environment*, vol.8 (3-4), pp.204-222.

Stern, P.C., Dietz, T. and Guagnano, G.A. (1998), 'A brief inventory of values', *Educational and Psychological Measurement*, vol.58 (6), pp.984-1001.

Stroh, M. (1998), Book Review of Noorman, K.J. and Uiterkamp, A.J.M. (eds), *Green Households? Domestic Consumers, Environment and Sustainability*, in *ECOS*, vol.19 (1), p. 109.

Tabachnik, B.G. and Fidell, L.S. (1996), *Using Multivariate Statistics*, Harper Collins, New York (Second Edition).

Tarrant, M.A. and Cordell, H.K. (1997), 'The effect of respondent characteristics on general environmental attitude-behavior correspondence', *Environment and Behavior*, vol.29 (5), pp.618-637.

Taylor, B. (1997), 'Green in word...', in Jowell, R., Curtice, J., Park, A., Brook, L., Thomson, K. and Bryson, C. (eds), *British Social Attitudes: the Fourteenth Report*, Ashgate, Aldershot, pp.111-136.

Taylor, S. and Todd, P. (1997), 'Understanding the determinants of consumer composting behavior', *Journal of Applied Social Psychology*, vol.27 (7), pp.602-628.

Thogersen, J. (1994), 'A model of recycling behaviour with evidence from Danish source separation programmes', *Journal of Research in Marketing*, vol.11 (1), pp.145-163.

Thogersen, J. (1996), 'Recycling and morality: a critical review of the literature', *Environment and Behavior*, vol.28 (4), pp.536-558.

Thompson, S.C.G. and Barton, M.A. (1994), 'Ecocentric and anthropocentric attitudes toward the environment', *Journal of Environmental Psychology*, vol.14 pp.149-157.

Thurstone, L.L. and Chave, E.J. (1929), *The Measurement of Attitude: A Psychological Method for Measuring Attitude Towards the Church*, University of Chicago Press, Chicago.

Tucker, L.R. (1978), 'The environmentally concerned citizen', *Environment and Behavior*, vol.10 (3), pp.389-418.

Tucker, P. (1999a), 'A survey of attitudes and barriers to kerbside recycling' *Environmental and Waste Management*, vol.2 (1), pp.55-62.

Tucker, P. (1999b), 'Normative influences in household recycling', *Journal of Environmental Planning and Management*, vol.42 (1), pp.63-82.

Turner, B.S. (1993), 'Contemporary problems in the theory of citizenship', in Turner, B.S. (ed.), *Citizenship and Social Theory*, SAGE, London, pp.1-18.

Turner, R.K. (1998), 'Household metabolism and sustainability', in Noorman, K.J. and Uiterkamp, A.J.M (eds), *Green Households: Domestic Consumers, Environment and Sustainability*, Earthscan, London, pp.1-6.

United Kingdom Election Statistics Database (2000), accessed 24th March 2000, via: www.ourwolrd.compuserve.com/homepages/zobbel/ge97_r02.htm.

United Nations Conference on Environment and Development (UNCED), (1992), *Agenda 21 - Action Plan for the Next Century*, endorsed at UNCED, UNCED, Rio de Janeiro.

Van Liere, K.D. and Dunlap, R.E. (1978), 'Moral norms and environmental behaviour: an application of Schwartz's norm-activation model to yard burning', *Journal of Applied Social Psychology*, vol.8 (2), pp.174-188.

Van Liere, K.D. and Dunlap, R.E. (1980), 'The social bases of environmental concern: a review of hypotheses, explanations and empirical evidence', *Public Opinion Quarterly*, vol.44, pp.181-197.

Van Liere, K.D. and Dunlap, R.E. (1981), 'Environmental concern: does it make a difference how it's measured?', *Environment and Behavior*, vol.13 (6), pp.651-676.

Van Steenbergen, B. (1994), 'Towards a global ecological citizen', in van Steenbergen, B. (ed.), *The Condition of Citizenship*, Sage, London, pp.141-152.

Vining, J. and Ebreo, A. (1990), 'What makes a recycler? A comparison of recyclers and nonrecyclers', *Environment and Behavior*, vol.22 (1), pp.55-73.

Vining, J. and Ebreo, A. (1992), 'Predicting recycling behaviour from global and specific environmental attitudes and changes in recycling opportunities', *Journal of Applied Social Psychology*, vol.22 (20), pp.1580-1607.

Waks, L.J. (1996), 'Environmental claims and citizen rights', *Environmental Ethics*, vol.18 (2), pp.133-148.

Wall, G. (1995), 'General versus specific concern: a western Canadian case', *Environment and Behavior*, vol.27 (3), pp.294-316.

Warner, L.G. and DeFleur, M.L. (1969), 'Attitude as an interactional concept: social constraint and social distance as intervening variables between attitudes and actions', *American Sociological Review*, vol.34 (2), pp.153-169.

Warren, N. and Jahoda, M. (1973), 'Introduction' in Warren, N. and Jahoda, M. (eds), *Attitudes*, Penguin, Harmondsworth, pp.9-15.

Warriner, G.K., McDougall, G.H. and Claxton, J.D. (1984), 'Any data or none at all? Living with inaccuracies in self-reports of residential energy consumption', *Environment and Behavior*, vol.16 (4), pp.513-526.

Watts, B.M. and Probert, E.J. (1999), 'Variation in public participation in recycling: a case study in Swansea', *Environmental and Waste Management*, vol.2 (2), pp.99-112.

Weigel, R.H. (1977), 'Ideological and demographic correlates of proecology behavior', *Journal of Social Psychology*, vol.103 pp.39-47.

Weigel, R.H. and Amsterdam, J.T. (1976), 'The effect of behavior relevant information on attitude-behaviour consistency', *Journal of Social Psychology*, vol.98 pp.247-251.

Weigel, R and Weigel, J. (1978), 'Environmental concern: the development of a measure', *Environment and Behavior*, vol.10 (1), pp.3-15.

Werner, C.M. and Makela, E. (1998), 'Motivations and behaviours that support recycling', *Journal of Environmental Psychology*, vol.18 pp.373-386.

Werner, C.M., Turner, J., Shipman, K., Twitchell, F.S., Dickson, B.R., Bruschke, G.V. and von Bismarck, W.B. (1995), 'Commitment, Behaviour, and Attitude change: An analysis of voluntary recycling', *Journal of Environmental Psychology*, vol.15 (3), pp. 197-208.

West, R. (1991), *Computing for Psychologists: Statistical Analysis Using SPSS and MINITAB*, Harwood, Chur.

Wicker, A.W. (1969), 'Attitudes versus actions: the relationship of verbal and overt behavioral responses to attitude objects', *Journal of Social Issues*, vol.25 (4), pp.41-78.

Widegren, O. (1998), 'The New Environmental Paradigm and personal norms', *Environment and Behavior*, vol.30 (1), pp.75-100.

Wilbanks, T.J. (1994), '"Sustainable development" in geographic perspective', *Association of American Geographers*, vol.84 (4), pp.541-556.

Witherspoon, S. and Martin, J. (1992), 'What do we mean by green?', in Jowell, R., Brook, L., Prior, G. and Taylor, B. (eds), *British Social Attitudes: The Ninth Report*, Dartmouth, Aldershot, pp.1-26.

Witmer, J.F. and Geller, E.S. (1976), 'Facilitating paper recycling: effects of prompts, raffles, and contests', *Journal of Applied Behavior Analysis*, vol.9 (3), 315-322.

World Commission on Environment and Development (WCED), (1987), *Our Common Future*, Oxford University Press, Oxford.

Young, J. (1990), *Post Environmentalism*, Belhaven, London.

Young, S.C. (1995), 'Participation - out with the old, in with the new?', *Town and Country Planning*, vol.64 (4), pp.110-112.

Index